CLASSICAL MECHANICS

Gary Gladding   Mats Selen   Tim Stelzer

University of Illinois at Urbana-Champaign

w. h. freeman   New York
Macmillan Learning

System Development: Michel Herquet, Geoffrey Piroux, and Michael L. Scott
Book Design, Illustrations, and Animations: Michael L. Scott

Publisher: Katherine Parker
Acquisitions Editor: Alicia Brady
Marketing Manager: Maureen Rachford
Marketing Assistant: Cate McCaffery
Managing Editor: Lisa Kinne
Project Editor: Jodi Isman
Director of Design, Content Management: Diana Blume
Cover Designer: Cambraia F. Fernandes
Print and Binding: Mercury Print Productions

ISBN 13: 978-1-319-06651-2
ISBN 10: 1-319-06651-8

Third Printing

Macmillan Learning
W.H. Freeman and Company
One New York Plaza
Suite 4500
New York, NY 10004-1562

www.flipitphysics.com

CONTENTS IN BRIEF

APPENDICES

CONTENTS

PART I   LINEAR DYNAMICS

1   ONE-DIMENSIONAL KINEMATICS / 1

2   TWO-DIMENSIONAL KINEMATICS / 13

3   RELATIVE AND CIRCULAR MOTION / 27

4   NEWTON'S LAWS / 37

10   CENTER OF MASS / 107

11   CONSERVATION OF MOMENTUM / 121

12   ELASTIC COLLISIONS / 133

13   COLLISIONS, IMPULSE AND REFERENCE FRAMES / 141

PART III   ROTATIONAL DYNAMICS

28 HEAT AND TEMPERATURE / 305

29 IDEAL GAS / 316

30 EQUIPARTITION, HEAT CAPACITY AND CONDUCTION / 326

31 HEAT ENGINES / 334

PREFACE

Welcome to FlipItPhysics, a ground-breaking learning environment for the calculus-based physics course.

STUDENTS

The textbook you hold in your hands is a companion to the online materials that precede and follow each lecture. Its streamlined, readable narrative emphasizes essential concepts and problem-solving strategies.

The textbook is to be used in tandem with the FlipItPhysics system:

1) **PreLectures** are animated, narrated introductions to the core concepts. Each PreLecture is approximately 20 minutes in length and includes embedded concept questions that allow students to check their understanding along the way.

2) Completion of the PreLecture unlocks a series of multiple choice and free response questions that gauge comprehension of the material. Results provide important feedback that instructors can use to tailor lectures to meet their students' needs.

3) With more useful preparation before class, the **Lecture** is transformed, becoming a place where students and instructors can interact, build on common understanding, and begin to transform facts into true knowledge.

4) Finally, FlipItPhysics offers a series of print and online **Problems and Interactive Tutorials**, which provide problem-solving models and include sophisticated, answer-specific feedback.

INSTRUCTORS

FlipItPhysics allows instructors to take advantage of a better-prepared student audience. It provides valuable feedback and tools, which facilitate the incorporation of proven just-in-time teaching methods and peer-instruction elements. Instructors have access to a wealth of data showing students' interaction with the course material, from the number of attempts on a given problem to the amount of time spent watching a PreLecture.

A robust suite of instructor resources makes FlipItPhysics implementation straightforward and flexible:

> **Lecture PowerPoints** will help make your transition to the FlipItPhysics system easy and efficient. Slides correspond directly with PreLecture and CheckPoint

content, and include designated spots for student data, a detailed notes section, and embedded clicker questions that can be used for classroom response.

Art Images and **Tables** are available to instructors in JPEG or PPT format.

Worked Solutions for all **Standard Exercises** are available to adopters.

The functionality of FlipItPhysics is intuitive and practical. Course setup is accomplished in just a few short steps. The calendar allows for drag-and-drop adjustment of course assignments. The gradebook monitors individual and class performance, and exports scores to standard campus course management systems.

Extensive and intuitive system support is an integral part of the FlipItPhysics system. Users have a comprehensive array of support options, from informative "sidebar" guidance that appears next to major functions, to a help-menu with detailed text-based and multimedia based instructions. Technical support can be reached by phone.

1-800-936-6899

Monday-Thursday: 7:00 a.m. to 3:00 a.m. EST
Friday: 7:00 a.m. to 11:00 p.m. EST
Saturday: 11:30 a.m. to 8:00 p.m. EST
Sunday: 11:30 a.m. to 11:00 p.m. EST

Our Approach

The FlipItPhysics approach is grounded in physics education research. The system was developed for over 10 years at the University of Illinois at Urbana-Champaign, and has been tested by thousands of student users and instructors at more than 75 institutions. Published research results show that students learn more when using FlipItPhysics and have a more positive outlook on physics and the lecture:

Stelzer, T., Gladding, G., Mestre, J.P., Brookes, D.T. (2009). Comparing the efficacy of multimedia modules with traditional textbooks for learning introductory physics content. *American Journal of Physics*, 77(2), 184-190.

Stelzer, T., Brookes, D.T., Gladding, G., Mestre, J.P. (2010). Impact of multimedia learning modules on an introductory course on electricity and magnetism. *American Journal of Physics*, 77(7), 755-759.

Chen, Z., Stelzer, T., Gladding, G. (2010). Using multimedia modules to better prepare students for introductory physics lectures. *Physics Review Special Topics – Physics Education Research*, 6, 010108, 1-5.

Sadaghiani, Homeyra R. (2011). Using multimedia learning modules in a hybrid-online course in electricity and magnetism. *Physics Review Special Topics – Physics Education Research*, 7, 010102, 1-7.

ABOUT THE AUTHORS

GARY GLADDING

Professor Gary Gladding, a high energy experimentalist, joined the Department of Physics at Illinois as a research associate after receiving his Ph.D. from Harvard University in 1971. He became assistant professor in 1973 and has, since 1985, been a full professor. He has performed experiments at CERN, Fermilab, the Stanford Linear Accelerator Center, and the Cornell Electron Storage Ring. He served as Associate Head for Undergraduate Programs for 13 years. He was named a Fellow of the American Physical Society for his contributions to the improvement of large enrollment introductory physics courses.

Since 1996, Professor Gladding has led the faculty group responsible for the success of the massive curriculum revision that has transformed the introductory physics curriculum at Illinois. This effort has involved more than 50 faculty and improved physics instruction for more than 25,000 science and engineering undergraduate students. He has shifted his research focus over the last ten years to physics education research (PER) and currently leads the PER group. He is also heavily involved in preparing at-risk students for success in physics coursework through the development of Physics 100. Professor Gladding was also a key player in the creation and development of i>clicker™.

MATS SELEN

Professor Mats Selen received his bachelor's degree in physics from the University of Guelph (1982), an M.Sc. in physics from Guelph (1983), and an M.A. in physics from Princeton University (1985). He received his Ph.D. in physics from Princeton (1989). He was a research associate at the Laboratory for Elementary Particle Physics (LEPP) at Cornell University from 1989 to 1993. He joined the Department of Physics at Illinois in 1993 as an assistant professor and, since 2001, has been a full professor. He was named a Fellow of the American Physical Society in 2006 for his contributions to particle physics.

Since arriving at Illinois, he has been a prime mover behind the massive curriculum revision of the calculus-based introductory physics courses (Physics 211-214), and he was the first lecturer in the new sequence. He created an undergraduate "discovery" course where freshmen create their own physics demonstrations, and developed the Physics Van Outreach program, in which physicists visit elementary schools to share enthusiasm for science. Professor Selen played a key role in the development of i>clicker™.

TIMOTHY J. STELZER

Professor Timothy Stelzer received his bachelor's degree in physics from St. John's University (1988) and his Ph.D. in physics from the University of Wisconsin-Madison (1993). After working as a senior research assistant in the Center for Particle Theory at Durham University (UK), he joined the Department of Physics at the University of Illinois as a postdoctoral research associate in 1995. Active in theoretical high-energy physics, he is currently a research associate professor at the university.

Professor Stelzer has been heavily involved with the Physics Education Group at Illinois, where he has led the development and implementation of tools for assessing the effectiveness of educational innovations in the introductory courses and expanding the use of Web technology in physics pedagogy. He was instrumental in the development of the i>clicker™ and is a regular on the University's "Incomplete List of Teachers Ranked as Excellent by Their Students." He was named University of Illinois Distinguished Teacher-Scholar in 2009.

ABOUT THE DEVELOPERS

MICHEL HERQUET

Michel received a M.Sc. in Theoretical Physics from the University of Mons in 2003 and a Ph.D. in Particle Physics from the University of Louvain in 2008. He also worked as a postdoctoral researcher at the Nikhef Institute in Amsterdam between 2008 and 2010. His research work was focused on phenomenology at CERN experiments and Monte Carlo simulations, in particular in the context of the hunt for the Higgs boson. He is one of the authors of MadGraph 4 and 5, two highly successful simulation packages totalling over 2000 citations. In 2010, he joined McKinsey & Company, a global management consulting firm, as a junior associate consultant and worked on projects in strategy and marketing/sales for clients in various industries. One year later, he helped in founding Novapta Consulting. He is also invited lecturer at the Louvain University, teaching Physics both at elementary and advanced levels.

GEOFFROY PIROUX

Geoffroy obtained a M.Sc. in Physics from the University of Namur in 1999 and a Ph.D. in Mathematical Physics from the University of Louvain in 2006. In parallel to his research, he studied Philosophy and obtained a B.A. degree in 2003.

His main research work focused on conformal field theories and their links with two dimensional critical dynamic models such as the sandpile model. In 2006, he joined Sopra Banking Software and worked in the product department as solution expert in the loans, treasury and risk management domains. He was also the manager of the compliance team that ensures the compatibility of the solution regarding regulatory, legal and third party products evolutions. Geoffroy is still involved in the academic field as invited lecturer at the University of Louvain where he teaches Physics for M.S. Business Engineering students.

MICHAEL L. SCOTT

Michael Scott received his bachelor's degree in physics and mathematics from the University of Indianapolis (2000) and both an M.S. in physics (2002) and a Ph.D. in physics (2008) from the University of Illinois at Urbana-Champaign. Dr. Scott's work at Illinois was in the field of physics education research (PER), with a focus on the effectiveness of multiple-choice exams in large introductory physics courses and on how explicit reflection can enhance physics learning. Michael received several teaching awards as a teaching assistant at Illinois, including the Scott Anderson Award (2002) and the AAPT Outstanding Teaching Assistant Award (2003) given by the Deparment of Physics. He also appeared numerous times on the University's "Incomplete List of Teachers Ranked as Excellent by Their Students." He is also a founding member of Novapta Consulting.

As a member of the smartPhysics team, Michael leads the development of the creative media, which includes animating the PreLectures and Problem-Solving Tutorials in Adobe Flash®. His responsibilities also include website design, textbook design, and illustrations.

COURSE OVERVIEW

The framework we will adopt in this course will be that introduced by Isaac Newton in the 17th century. This framework remained the standard in science until the 20th century when fundamental changes were needed to describe the complete nature of space, time, and matter. In particular, the theory of special relativity proposed a constant speed of light that led to a reformulation of the nature of space and time. The theory of quantum mechanics was created to describe the interactions of elementary particles, such as electrons and photons, leading to a description of matter that included both particle and wavelike aspects. In this course, we will restrict ourselves to describing macroscopic objects moving at relative velocities that are small with respect to the speed of light, so that Newtonian mechanics is all we need to accurately describe the physics.

In particular, we will present Newton's laws which introduce the new concepts of force and mass that are needed to describe the actions of macroscopic objects. Indeed, Newton's laws establish the mechanical world view that forms the basis for the scientific revolution of the 17th century. In particular, he introduced a universal force of gravitation that he claimed applied to all objects having mass. He then demonstrated that he could relate the motion of the Moon in its orbit about the Earth to the falling of an apple to the ground here on Earth. The deep significance of this demonstration was that, for the first time, a connection was made between the motions of ordinary things on the Earth and the motions of heavenly bodies. Prior to Newton, the Heavens and the Earth were treated entirely separately and differently. Newton showed that the laws that apply here on Earth extend to the Heavens.

We will then introduce other important quantities such as energy, momentum, and angular momentum that are commonly conserved in a variety of situations. We'll close with a brief study of oscillatory motions, wave motions, and fluids.

UNIT

1

ONE-DIMENSIONAL KINEMATICS

1.1 Overview

This course is concerned with classical mechanics, the study of the forces and motions of macroscopic objects. We will begin with a study of **kinematics**, the description of motion, without regard to its cause. In particular, we will define the concepts of *displacement*, *velocity* and *acceleration* that are needed to describe motion. We will initially restrict ourselves to motions in one dimension. We will use these definitions to demonstrate how to obtain the *change in position* from the velocity and the *change in velocity* from the acceleration. We will close this unit with a discussion of an example of a particular motion, that of constant acceleration.

1.2 Displacement and Average Velocity

To discuss motion in classical physics, we begin with two quantities, *displacement* and *velocity*. These quantities are not unfamiliar to you; We're sure you already have a working knowledge of the relationship between displacement and velocity. If it takes you three hours to walk six miles, you can Figure out that your average velocity during that walk was 2 miles/hr by simply dividing the distance you walked by the time it took.

In this course, you will find that many of the words that represent the quantities of physics will be very familiar words. It is important to note, however, that these words all have very precise meanings in physics, whereas in everyday language, these words are often used to mean many related, but different, things. Therefore, it is important that we start right away with careful definitions of our terms. Most often, these definitions will obtain their precision through their expression in terms of mathematics.

To illustrate this point, we introduce an argument made by the Greek philosopher Zeno to prove that it is impossible to move from some point A to another point B. His argument goes as follows: clearly before we can move to point B, we need first to move to point C which is halfway between points A and B. Sounds true enough, however, this argument can be repeated *ad infinitum*; i.e., once at C, we would need to move first to point D which is halfway between points C and B. You get the drift. We will need to make an infinite number of moves to get to point B.

What is Zeno's point? It certainly is *not* to prove that motion is impossible; we all know that is not true. In fact, the reason that these arguments are called "paradoxes" is that what seems to be a reasonable argument leads to a conclusion that we know is false. Zeno initiated these arguments as ways to investigate the nature of space and time.

How do we resolve these paradoxes? Clearly the problem lies with the notion of infinity. Mathematics can help. We know, for example, that an infinite series can have a well-defined sum. This sum is defined in terms of a limit which is the key concept of calculus. Indeed, we will soon find that the use of calculus will be central to the definition of velocity. For now, though, we'll begin by defining the **average velocity** of an object within some time interval Δt to be equal to Δx, the distance it has traveled during that period divided by the time it takes.

$$v \equiv \frac{\Delta x}{\Delta t}$$

We can represent this definition graphically as shown in Figure 1.1. On the vertical axis we plot the displacement x, which is defined to be the distance traveled from some fixed origin, while on the horizontal axis, we plot the time t, from some fixed time defined to be $t = 0$. If we choose some time interval defined by the times t_i and t_f, we see the corresponding displacements x_i and x_f and that the average velocity is just the slope of the line connecting the initial and final points on the graph.

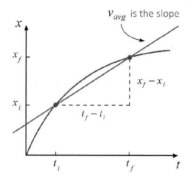

FIGURE 1.1 A plot of the displacement *x* as a function of time *t*. The *average velocity* for a time interval (t_i , t_f) is illustrated as the slope of the line connecting the points on the curve at those two times.

To illustrate the use of the average velocity, suppose someone calls your cell phone while you are driving and asks "What's up?" You might say, "I left town half an hour ago and I'm heading east on the interstate." If you wanted to be more specific you might say, "I am 35 miles east of town, and I'm driving east at an average speed of 65 miles per hour." From this information your friend could do a simple mental calculation to predict that one hour from now you will be another 65 miles farther east, which would be about a hundred miles east of town.

What your friend really did to make this estimate was to solve the following kinematic equation in her head.

$$x(\Delta t) = x_o + v_{avg} \cdot \Delta t$$

The translation of this equation into English is that your position at a time Δt after you start is just equal to your starting position, call it x_o , plus the additional distance you went during time Δt. This last piece is just your average velocity multiplied by the elapsed time. This last calculation assumes that your average velocity does not change in the next hour. We will discuss how to describe motion in which your velocity *does change* in the next section

Before we proceed any further, a remark about units is in order. Although the velocity used in this example was given in miles per hour, we will adopt for the most part in this course, the SI system of units in which velocity is measured in meters/second. Converting between these units is easy; we simply have to multiply by "one" until the units are right. For example in this case, we can multiply 65 miles/hour by 1,609 meters/mile by (1/3,600) hours/second to obtain the result that 65 mph is equivalent to 29 m/s.

1.3   Instantaneous Velocity

In the last example, we calculated the predicted distance the car would go during a specified period of time, assuming that the average velocity during that time did not change. You know this assumption is not always true; sometimes you may speed up to pass a car, resulting in an increased average velocity, or you may have to slow down due to traffic, resulting in a decreased average velocity.

Therefore, to discuss all kinds of motion, we will need the ability to figure out both the displacement and the velocity for any instant in time, not just the average over some time interval.

We can visualize the procedure for finding the instantaneous velocity by starting with the displacement versus time plot shown in Figure 1.1 and then bringing the final and initial times closer and closer together until they are infinitesimally close together. As we do this the line connecting the points becomes the tangent to the curve as shown in Figure 1.2! In other words, the **instantaneous velocity** at some time t is just the slope of the tangent to the x versus t curve at that point. The slope of this tangent line is exactly equal to the derivative dx/dt at that time!

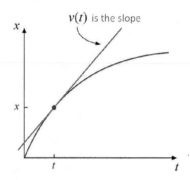

FIGURE 1.2 A plot of the displacement x as a function of time t. The *instantaneous velocity* at time t is illustrated as the slope of the tangent to the curve at time t.

We can now see the simple relationship between *displacement* and *velocity*: The instantaneous velocity at a particular time t is defined to be the time derivative of the displacement at that time.

$$v \equiv \frac{dx}{dt}$$

We can construct a graph of the instantaneous velocity as a function of time by finding the slope of the corresponding x versus t graph at each time t as shown in Figure 1.3.

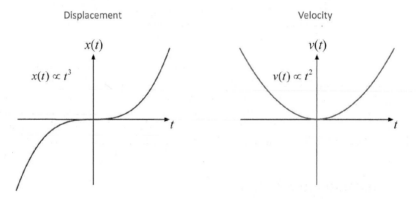

FIGURE 1.3 The instantaneous velocity at any time is obtained by differentiating the displacement at that time.

This relationship is the definition of instantaneous velocity and is *always* true, no-matter how strangely the displacement may be changing with time. We were led to this definition by a natural refinement of the concept of the average velocity, and we have discovered, as predicted, that the calculus (in this case, the derivative) is needed to carefully define this kinematic quantity.

1.4   Position from Velocity

We've just defined the instantaneous velocity at time t as the time derivative of the displacement at time t. Therefore, if we know the displacement as a function of time for some object, we can calculate its velocity at any time by simply evaluating the derivative of the displacement function at that time.

Suppose, on the other hand, that we know the velocity as a function of time; what can we say then about the displacement at any time? It seems like we should be able to use the inverse operation to go the other way–to evaluate the integral of the v versus t graph to find the displacement as a function of time.

We know the integral can be represented graphically as the area under the curve. Therefore, we expect the displacement to be related to the area under the v versus t graph. We can verify this expectation for the special case of motion with a constant velocity as shown in Figure 1.4. In this case the area under the curve from 0 to time t is simply equal to the magnitude of the velocity times the time t. Note that the integral needed to find the displacement at time t is the definite integral from $t = 0$ to $t = t$.

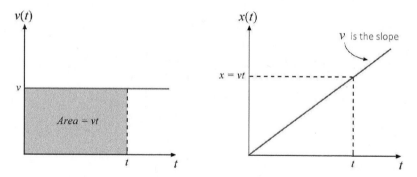

FIGURE 1.4 For motion at constant velocity, the displacement as a function of time can be obtained by integrating the constant velocity over time to obtain a displacement that changes linearly with time.

To obtain the general equation that determines the displacement from the velocity, we actually need to be a little more careful. What we've really shown so far is that the *change* in displacement during a time interval (for example, t_i to t_f) is equal to the integral of the velocity between these initial and final times. We have no way of knowing where the particle was at any particular time, say at $t = 0$. The velocity tells us how the displacement changes; it can't tell us where it started from. That information must be given to us independently. Consequently, we write the general expression in terms of the definite integral of the velocity from t_i to t_f and the value of the velocity at t_i.

$$x(t_f) - x(t_i) = \int_{t_i}^{t_f} v(t)\, dt$$

This expression is completely general and will work for any velocity function.

1.5 Acceleration

There is one more important kinematic quantity that we need to discuss. Namely, just as velocity tells us how fast the displacement is changing, **acceleration** tells us how fast the velocity is changing. In other words, acceleration is the time rate of change of velocity; acceleration is the measure of how many meters per second the velocity changes in a second. The units of acceleration are therefore meters per second per second.

$$a \equiv \frac{dv}{dt}$$

Figure 1.5 shows a plot of the velocity of an object as a function of time. The value of the acceleration at any given time is just equal to the slope of this curve at that time.

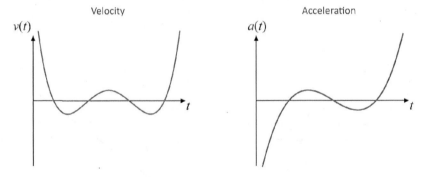

FIGURE 1.5 The acceleration of an object at any time is obtained by differentiating its velocity at that time.

In the same way that the change in displacement can be found by integrating the velocity, the change in velocity can be found by integrating the acceleration.

Before going any further, we must issue here a warning about a common confusion between everyday language and precise kinematic definitions. The evil word here is *deceleration.* We're sure practically all of you associate decelerating with "slowing down." We have just defined acceleration as the time rate of change of velocity. The velocity is a *signed number* as is the acceleration. The sign of the velocity indicates its direction, either forward (positive) or backward (negative). If a car is moving in the positive direction and slowing down, then its acceleration is negative. However, if the car is moving in the negative direction and slowing down, then its acceleration is *positive!*

Therefore, the concept of "slowing down" is not the same as that of "negative acceleration." The safest way to proceed here is to just not use the word deceleration when dealing with kinematics problems. Accelerations are either positive or negative, depending on whether the velocity, a signed number, is increasing or decreasing.

1.6 Constant Acceleration

The equations we presented in the last section for the acceleration as the time rate of change of the velocity and the change in velocity as the integral of the acceleration are totally general; they are true always!

We'd like now to use these general equations to derive the specific equations that hold for a special, but important case, namely that of *motion at constant acceleration.*

We start with the defining property, that the acceleration is a constant. We can integrate this constant acceleration to find the change in velocity. The result of this integration is that the velocity at any given time is simply equal to the initial velocity plus the acceleration multiplied by the elapsed time.

$$v(t_f) = v(t_i) + a \cdot (t_f - t_i)$$

This equation is often written in a more compact form.

$$v = v_o + at$$

In writing this equation, we denote the velocity at the initial time by v_o, and the variable t really means the elapsed time $t_f - t_i$. We see that the velocity changes linearly with time, as it must since the acceleration is constant.

Now that we have the velocity as a function of time, we can integrate once again to find the displacement as a function of time.

$$x = x_o + \int_0^t (v_o + at)\, dt$$

In this case we see that the displacement changes quadratically with time.

$$x = x_o + v_o t + \frac{1}{2} at^2$$

We have now obtained expressions for the velocity and displacement as a function of time for the special case of motion at constant acceleration. We can eliminate the time from these equations, for example by solving for t in the velocity equation and substituting that expression back into the displacement equation to obtain a new expression that directly relates the velocity to the displacement.

$$2a(x - x_o) = v^2 - v_o^2$$

In particular, we see that the displacement increases as the square of the velocity.

Main Points

Definitions of Kinematic Quantities

Displacement $x(t)$

Velocity $v(t) \equiv \dfrac{dx(t)}{dt}$ Velocity is the time rate of change of displacement.

Acceleration $a(t) \equiv \dfrac{dv(t)}{dt}$ Acceleration is the time rate of change of velocity.

Obtaining Displacement and Velocity from Acceleration

Displacement $x(t_f) - x(t_i) = \displaystyle\int_{t_i}^{t_f} v(t)\,dt$ Displacement is the integral of the velocity over time.

Velocity $v(t_f) - v(t_i) = \displaystyle\int_{t_i}^{t_f} a(t)\,dt$ Velocity is the integral of the acceleration over time.

Special Case: Motion with Constant Acceleration

The displacement is obtained by integrating the velocity over time.

The velocity is obtained by integratng the constant acceleration over time.

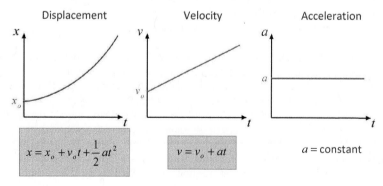

Displacement	Velocity	Acceleration

$$x = x_o + v_o t + \frac{1}{2}at^2 \qquad v = v_o + at \qquad a = \text{constant}$$

PROBLEMS

1. Car Ride: Two cars start from rest at a red stop light. When the light turns green, both cars accelerate forward. The blue car accelerates uniformly at a rate of 3.3 m/s^2 for 3 s. It then continues at a constant speed for 12.8 s, before applying the brakes such that the car's speed decreases uniformly coming to rest 152 m from where it started. The yellow car accelerates uniformly for the entire distance, finally catching the blue car just as the blue car comes to a stop. (a) How fast is the blue car going 2.1 s after it starts? (b) How fast is the blue car going 9.9 s after it starts? (c) How far does the blue car travel before its brakes are applied to slow down? (d) What is the acceleration of the blue car once the brakes are applied? (e) What is the total time the blue car is moving? (f) What is the acceleration of the yellow car?

2. Tortoise and Hare: A tortoise and hare start from rest and have a race. As the race begins, both accelerate forward. The hare accelerates uniformly at a rate of 1.6 m/s^2 for 4.3 s. It then continues at a constant speed for 12.4 s, before getting tired and slowing down with constant acceleration coming to rest 110 m from where it started. The tortoise accelerates uniformly for the entire distance, finally catching the hare just as the hare comes to a stop. (a) How fast is the hare going 2.6 s after it starts? (b) How fast is the hare going 10.9 s after it starts? (c) How far does the hare travel before it begins to slow down? (d) What is the acceleration of the hare once it begins to slow down? (e) What is the total time the hare is moving? (f) What is the acceleration of the tortoise?

3. Two Thrown Balls: A blue ball is thrown upward with an initial speed of 21.2 m/s, from a height of 0.9 m above the ground. 2.6 s after the blue ball is thrown, a red ball is thrown down with an initial speed of 10.6 m/s from a height of 25.4 m above the ground. The force of gravity due to the Earth results in the balls each having a constant downward acceleration of 9.81 m/s^2. (a) What is the speed of the blue ball when it reaches its maximum height? (b) How long does it take the blue ball to reach its maximum height? (c) What is the maximum height the blue ball reaches? (d) What is the height of the red ball 3.38 seconds after the blue ball is thrown? (e) How long after the blue ball is thrown are the two balls in the air at the same height? (f) Which statement is true about the blue ball after it has reached its maximum height and is falling back down?
  (i)   The acceleration is positive, and it is speeding up.
  (ii)   The acceleration is negative, and it is speeding up.
  (iii)   The acceleration is positive, and it is slowing down.
  (iv)   The acceleration is negative, and it is slowing down.

4. I-74 (INTERACTIVE EXAMPLE): Anna is driving from Champaign to Indianapolis on I-74. She passes the Prospect Ave. exit at noon and maintains a constant speed of 75 mph for the entire trip. Chuck is driving in the opposite direction. He passes the Brownsburg, IN exit at 12:30 pm and maintains a constant speed of 65 mph all the way to Champaign. Assume that the Brownsburg and Prospect exits are 105 miles apart and that the road is straight. How far from the Prospect Ave. exit do Anna and Chuck pass each other?

5. V versus T (INTERACTIVE EXAMPLE): Shown below is a graph of velocity versus time for a moving object. The object starts at position $x = 0$. What is the final position (as measured from $x = 0$) after it experiences the motion described by the graph, from $t = 0$ seconds to $t = 5$ s?

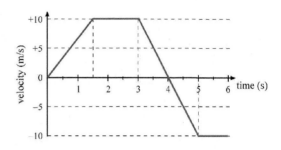

FIGURE 1.6 Problem 5

UNIT

2

VECTORS AND TWO-DIMENSIONAL KINEMATICS

2.1 Overview

We will begin by introducing the concept of vectors that will allow us to generalize what we learned last time in one dimension to two and three dimensions. In particular, we will define vector addition and subtraction and relate the component representation of a vector to the magnitude and direction representation.

We will then introduce one physics example, namely the description of free fall near the surface of the Earth as motion of constant acceleration in the vertical direction and motion at constant velocity in the horizontal direction. We will use these descriptions to calculate some properties of projectile motion. Finally, we will use the principle of superposition to relate the descriptions of projectile motion in two different reference frames.

2.2 Kinematic Definitions in Three Dimensions

To this point, we have restricted ourselves to discussions of motions in one dimension; we have defined velocity as dx/dt and acceleration as dv/dt. How do we generalize these definitions to more than one dimension? The generalization we make is most easily understood in terms of Cartesian coordinates. Figure 2.1(a) shows a Cartesian coordinate system, with the mutually orthogonal directions labeled x, y, and z. To identify a point P in this space, we can specify its three coordinates (x, y, z). These coordinates represent how far the point is from the origin in the x, y, and z directions. If we draw an arrow from the origin to the point, as shown in Figure 2.1(b), we can define this arrow as the **displacement vector** that locates the point; the coordinates (x, y, z) are called the components of the displacement vector in this system. With this definition of the displacement vector, it is natural to define the components of the **velocity vector** and the **acceleration vector** similarly.

$$v_x \equiv \frac{dx}{dt} \qquad v_y \equiv \frac{dy}{dt} \qquad v_z \equiv \frac{dz}{dt}$$

$$a_x \equiv \frac{dv_x}{dt} \qquad a_y \equiv \frac{dv_y}{dt} \qquad a_z \equiv \frac{dv_z}{dt}$$

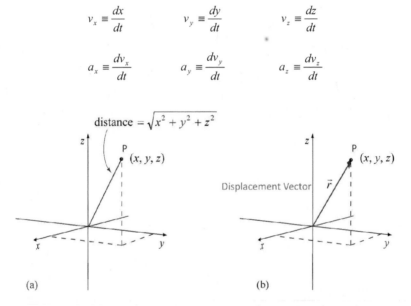

$$\text{distance} = \sqrt{x^2 + y^2 + z^2}$$

(a)

(b)

FIGURE 2.1 (a) A point P is specified in a Cartesian coordinate system by its components (x, y, z). (b) A point P is specified by its displacement vector $\vec{r}$ whose Cartesian coordinates are (x, y, z).

With these definitions, we can see that everything we did last time for one dimension (x) is just repeated for the other two dimensions $(y$ and $z)$. For example, we can immediately write down the equations for all the components for motion at constant acceleration.

$$v_x = v_{ox} + a_x t \qquad v_y = v_{oy} + a_y t \qquad v_z = v_{oz} + a_z t$$

$$x = x_o + v_{ox} t + \frac{1}{2} a_x t^2 \qquad y = y_o + v_{oy} t + \frac{1}{2} a_y t^2 \qquad z = z_o + v_{oz} t + \frac{1}{2} a_z t^2$$

We can formalize this generalization from one dimension to three dimensions by defining these kinematic quantities, displacement, velocity and acceleration as *vector quantities*. For example, we can write down a single equation for the velocity vector as a function of time for the special case of constant acceleration.

$$\vec{v} = \vec{v}_o + \vec{a}t$$

The "arrow" notation is apt here since it indicates that, like an arrow, a vector has both a *length* and a direction. The length of a vector is also called its magnitude and is often represented as the absolute value of the vector. This single *vector equation* is equivalent to the three scalar equations we wrote down earlier.

$$v_x = v_{ox} + a_x t \qquad\qquad v_y = v_{oy} + a_y t \qquad\qquad v_z = v_{oz} + a_z t$$

We have introduced these vectors in terms of one representation, their Cartesian components. In fact, you should think of these **vectors** as the primary object. They can have several different scalar component representations. In the next section, we will support this claim by introducing some important properties of vectors that we will use often in this course.

2.3   Vectors

You know how to perform many operations on scalar quantities. For example, you know how to add, subtract, multiply and divide numbers. You also know how to differentiate and integrate scalar functions. We can define similar operations for vectors.

For example, Figure 2.2 shows the procedure for defining the sum of two vectors $\vec{A}$ and $\vec{B}$. Namely, this sum is defined to be another vector $\vec{C}$ which is obtained from $\vec{A}$ and $\vec{B}$ using the following prescription: place the tail of vector $\vec{B}$ at the head of vector $\vec{A}$ and then draw the arrow from the tail of vector $\vec{A}$ to the head of vector $\vec{B}$. Note that the vector sum depends on the directions of the vectors as well as their magnitudes. For example, if you were to rotate vector $\vec{B}$ through some angle, its magnitude would not change, but both the direction and the magnitude of the vector sum $\vec{C}$ will change! Clearly the magnitude of the vector sum $\vec{C}$ is *not* equal to the sum of the magnitudes of vectors $\vec{A}$ and $\vec{B}$.

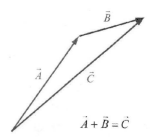

$$\vec{A} + \vec{B} = \vec{C}$$

FIGURE 2.2 The sum of two vectors $\vec{A}$ and $\vec{B}$ is defined to be another vector $\vec{C}$ formed by placing the tail of $\vec{B}$ at the head of $\vec{A}$ and drawing a vector from the tail of $\vec{A}$ to the head of $\vec{B}$.

You must be thinking that this is a pretty strange prescription to be given the name of something simple like addition. This prescription becomes clearer if we look at the Cartesian components of the vectors as shown in Figure 2.3.

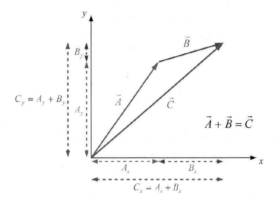

FIGURE 2.3 The components of the vector sum $\vec{C}$ are equal to the sum of the components of the vectors $\vec{A}$ and $\vec{B}$.

Aha, there is a method to this madness! It's clear that this definition of vector addition gives the result that the Cartesian components simply add!

$$C_x = A_x + B_x \qquad\qquad C_y = A_y + B_y$$

With this definition of vector addition, we see we can also write a general expression of any vector $\vec{A}$ in terms of its Cartesian components and the unit vectors in the (x, y, z) directions, as shown in Figure 2.4. Here we have used the fact that multiplying a vector by a scalar is the same as multiplying each of its components by the same scalar, which simply changes its length. Multiplying a vector by a negative scalar reverses its direction.

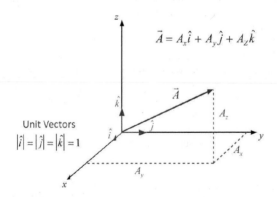

FIGURE 2.4 The vector $\vec{A}$ represented as the vector sum of the product of its components and the corresponding unit vector.

The Cartesian component representation of a vector is a common representation, but certainly not the only one. For example, a vector can also be specified in spherical components, in which the length of the vector and the angles describing its orientation are used to specify the vector as in Figure 2.5(a). In two dimensions, the orientation of a vector can be specified by the angle θ it makes with the x-axis as shown in Figure 2.5(b). Using trigonometry, we can determine completely the relation between the Cartesian components (A_x, A_y) and the polar components (A, θ).

In all cases, though, you should think of the vector itself as an object – the arrow. The different coordinate systems we invent are just different ways of describing this object in terms of scalar quantities.

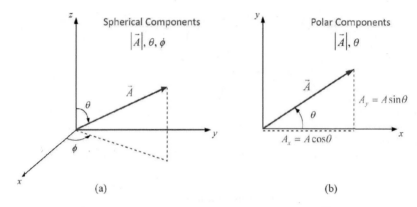

(a)   (b)

FIGURE 2.5 (a) The representation of a three-dimensional vector $\vec{A}$ in spherical coordinates. (b) The representation of a two-dimensional vector $\vec{A}$ in polar coordinates.

2.4   Example: Free Fall (Gravity)

We will now return to do some physics by considering an example of motion in three dimensions with constant acceleration, namely the throwing of a ball across a room. Once the ball leaves our hand the only force acting on it is gravity. We will learn more about gravity in a few units. For now the only thing we need to know is that near the surface of the Earth any object under the influence of just gravity (i.e., in free fall) will experience the same downward acceleration of 9.8 m/s^2. It is customary to refer to the magnitude of the acceleration of gravity as g. Figure 2.6 shows the familiar parabolic trajectory followed by the ball once it is in the air.

Before attempting to describe this motion using our new 3-D kinematics equations we need to define our coordinate system. It is customary to pick the y-axis to point vertically upward and the x-axis to point horizontally in the direction of the throw. With this choice, our kinematics equations simplify considerably as shown in Figure 2.7.

FIGURE 2.6 The parabolic trajectory followed by a thrown ball.

Since the acceleration is only in the $-y$ direction, a_x and a_z are zero. Therefore, the velocities in the x and z directions cannot change; the motion in these directions is just motion at constant velocity. Since we chose v_{oz} to be zero, we have no motion along the z direction at all. The motion of the ball will be restricted to the x-y plane; we have reduced

Motion with Constant Acceleration

$$a_x = 0 \qquad\qquad a_y = -g \qquad\qquad a_z = 0$$

$$v_x = v_{ox} \qquad\qquad v_y = v_{oy} - gt \qquad\qquad v_z = v_{oz}$$

$$x = x_o + v_{ox}t \qquad y = y_o + v_{oy}t - \frac{1}{2}gt^2 \qquad z = z_o$$

FIGURE 2.7 The kinematic equations above for a ball thrown across the room.

the problem to a two-dimensional problem. Indeed, we might as well choose the initial z position (z_o) to be zero which results in these simplified equations.

$$a_x = 0 \qquad\qquad v_x = v_{ox} \qquad\qquad x = x_o + v_{ox}t$$

$$a_y = -g \qquad\qquad v_y = v_{oy} - gt \qquad\qquad y = y_o + v_{oy}t - \frac{1}{2}gt^2$$

2.5 Example: Soccer Ball Kick

We will now use our knowledge of the motion of an object in free fall near the surface of the Earth to make a calculation. Suppose you kick a ball off the ground at an angle θ with an initial speed v_o. How far away will it land?

The equations we developed in the last section already reflect the fact that we have chosen the y-axis to be up and the x-axis to be in the direction of the kick. To make things even simpler, lets kick the ball at $t = 0$ and choose the origin of our coordinate system to be at the initial position of the ball so that $x_o = y_o = 0$. We now have equations for all three quantities that change as a function of time (x, y, and v_y).

$$x = v_{ox}t \qquad\qquad y = v_{oy}t - \frac{1}{2}gt^2 \qquad\qquad v_y = v_{oy} - gt$$

We want to determine the horizontal distance the ball travels before hitting the ground, call it D. Suppose the ball hits the ground at time $t = t_f$. The distance D then is really just the x position of the ball at time $t = t_f$. Therefore, we use the x equation to tell us that D is just equal to the product of the time t_f and the v_{ox}, the x component of the initial velocity.

$$x = v_{ox}t_f$$

Therefore, to determine D, we must first determine t_f and v_{ox}. How do we do that?

We can certainly determine v_{ox} from trigonometry. Namely, all we have to do is to decompose the initial velocity vector into its Cartesian components as shown in Figure 2.8. Since these components form the sides of a right triangle whose hypotenuse is equal to v_o, the magnitude of the initial velocity, we see that:

$$v_{ox} = v_o \cos\theta \qquad\qquad v_{oy} = v_o \sin\theta$$

Since both v_o and θ are given in the problem statement, we now know both v_{ox} and v_{oy} as well. The only remaining task is to determine t_f, the time the ball stays in the air. Since we have an equation

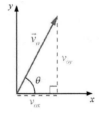

FIGURE 2.8 The decomposition of the initial velocity vector $\vec{v}_o$ into its components, v_{ox} and v_{oy}.

for y as a function of t we can just solve this equation for the times at which $y = 0$.

$$0 = v_{oy}t_f - \frac{1}{2}gt_f^2$$

Since this is a quadratic equation, we will find two solutions:

$$t_f = 0 \qquad\qquad t_f = \frac{2v_{oy}}{g}$$

These are the times at which the height of the ball was zero: one is when the ball was kicked ($t_f = 0$), and the other is when the ball landed ($t_f = 2v_{oy}/g$).

Finally, we just plug this last value for t_f in our equation to determine that the horizontal distance the ball travels in the air is proportional to the product of the x and y components of the initial velocity and inversely proportional to the acceleration due to gravity, g.

$$D = \frac{2v_{ox}v_{oy}}{g}$$

2.6   The Range Equation

When we solved the problem in the last section we found that the distance the ball travels was proportional to both v_{ox} and v_{oy}. It's always a good idea to check your results to see if they make sense.

If we increase v_{ox} the ball moves farther along the x direction in a given amount of time. If we increase v_{oy} the ball will be in the air longer and will travel farther for any given velocity in the x direction. This all makes sense, the catch is that for a given initial speed v_o, both v_{ox} and v_{oy} depend on the angle at which the ball is kicked. Therefore, an interesting question is: for what angle is this distance D a maximum? How would we go about making this calculation? The first step is to determine a general expression for the distance D in terms of the angle θ. We can then examine this expression to determine the angle that maximizes D.

We'll start with the expression for D that we obtained in the last section and write both v_{ox} and v_{oy} in terms of v_o and θ.

$$D = \frac{2v_o^2 \cos\theta\sin\theta}{g}$$

We see that D is proportional to the product of $\sin\theta$ and $\cos\theta$. We can simplify this expression a bit further by realizing the product of $\sin\theta$ and $\cos\theta$ is proportional to the $\sin(2\theta)$ to obtain the usual form of the **range equation**:

$$D = \frac{v_o^2 \sin 2\theta}{g}$$

Figure 2.9(a) shows a plot of the range D as a function of θ. We see that D reaches its maximum value when $\theta = 45°$. Figure 2.9(b) shows trajectories for different values of θ.

We see that the range for complementary angles is the same and that the maximum range is indeed obtained when $\theta = 45°$.

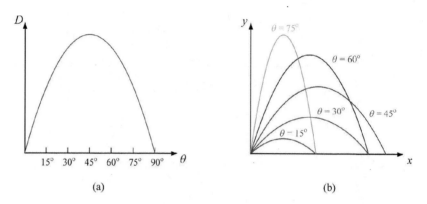

(a)   (b)

FIGURE 2.9 (a) A plot of the range D of a kicked soccer ball as a function of θ, the angle the ball's initial velocity vector makes with the horizontal. (b) Trajectories of the ball for different values of θ.

2.7 Superposition

We have already seen that the x and y equations of projectile motion are independent. In practical terms this means that the behavior of a projectile in the vertical direction is the same no matter how fast it is moving in the x direction, and vice versa. A bullet shot horizontally out of a gun will take the same amount of time to hit the level ground as a bullet dropped from the same height.

Therefore, we can consider the motion of the kicked ball to be the superposition of two simpler motions, the first being that of a ball moving vertically with constant acceleration, and the second being that of a ball moving horizontally with constant velocity. We can actually see this superposition if we simply consider a single motion as viewed from two different reference frames.

For example, if a man throws a ball vertically upward, we know the ball will go straight up and then straight down, all the time moving with constant acceleration of $9.8 \, \text{m/s}^2$ pointed downward. Now, suppose this man is sitting in a train while the train is moving with a constant speed past an observer at a station. What will he see? Well, if the train is really moving with constant speed, then the man on the train must see exactly what he saw before; the ball goes straight up and returns to his hand. The speed of the train makes no difference, as long as it's constant. What does the observer at the station see? He can't really see the same thing. Figure 2.10 shows the trajectory of the ball as seen by the observer on the ground. He does not see the ball go simply straight up and down!

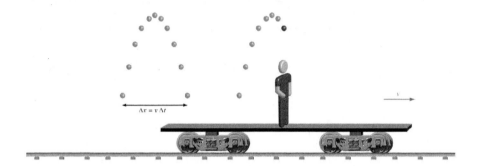

FIGURE 2.10 The trajectory of a ball thrown straight up by a man at rest on a train moving at constant velocity v with respect to an observer on the ground. In the time Δt it takes the ball to go straight up and down with respect to the man on the train, the train has traveled a distance $\Delta x = v \Delta t$. The ball is always directly above the man on the train and therefore appears to have the trajectory shown to the observer on the ground.

He sees the train moving so that the horizontal positions of the ball when it leaves the man's hand and when it returns are separated by the distance the train travels during the flight time of the ball. In fact, what he sees is the combined motions of constant acceleration in the vertical direction and constant velocity in the horizontal direction.

What he sees is exactly the trajectory a soccer ball would have if it were kicked with an initial velocity such that its vertical component were the initial velocity of the ball with respect to the man sitting on the train and its horizontal component were the velocity of the train with respect to the observer at the station.

The amazing conclusion we take away from this analysis is that projectile motion can be explained as simply free fall as viewed from a moving reference frame!

For example, in this case, we can predict what the man at the station will see by combining the information of what the man on the train sees with the known motion of the train. In particular, we can write a vector equation that relates the velocity of the ball as measured by the observer at the station to the velocity of the ball as measured by the man on the train.

$$\vec{v}_{ball,ground} = \vec{v}_{ball,train} + \vec{v}_{train,ground}$$

This vector equation relates observations in two different reference frames that are moving relative to each other and will be the topic of our next unit.

MAIN POINTS

Kinematic Quantites as Vector Quantities

Vectors have magnitude and direction.

Magnitude of Vector $\vec{v}$

$$|\vec{v}| = \sqrt{v_x^2 + v_y^2}$$

Displacement $\vec{x}$

Velocity $\vec{v} \equiv \dfrac{d\vec{x}}{dt}$

Acceleration $\vec{a} \equiv \dfrac{d\vec{v}}{dt}$

Vector Addition

The vector sum of two vectors $\vec{A}$ and $\vec{B}$ is another vector $\vec{C}$.

$$C_x = A_x + B_x$$

$$C_y = A_y + B_y$$

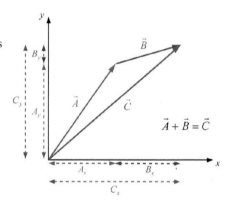

$$\vec{A} + \vec{B} = \vec{C}$$

Special Case: Projectile Motion

Projectile motion is the superposition of two independent motions:
1) **Horizontal**: constant velocity
2) **Vertical**: constant acceleration

Projectile motion can be understood simply as free fall viewed from a moving reference frame.

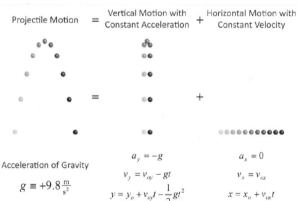

Projectile Motion = Vertical Motion with Constant Acceleration + Horizontal Motion with Constant Velocity

Acceleration of Gravity

$$g \equiv +9.8 \tfrac{\mathrm{m}}{\mathrm{s}^2}$$

$$a_y = -g$$
$$v_y = v_{oy} - gt$$
$$y = y_o + v_{oy}t - \frac{1}{2}gt^2$$

$$a_x = 0$$
$$v_x = v_{ox}$$
$$x = x_o + v_{ox}t$$

PROBLEMS

1. Catch: Julie throws a ball to her friend Sarah. The ball leaves Julie's hand a distance 1.5 m above the ground with an initial speed of 23 m/s at an angle of 26° with respect to the horizontal. Sarah catches the ball 1.5 m above the ground. (a) What is the horizontal component of the ball's velocity when it leaves Julie's hand? (b) What is the vertical component of the ball's velocity when it leaves Julie's hand? (c) What is the maximum height the ball goes above the ground? (d) What is the distance between the two girls? (e) After catching the ball, Sarah throws it back to Julie. The ball leaves Sarah's hand a distance of 1.5 m above the ground, and it is moving with a speed of 25 m/s when it reaches a maximum height of 9 m above the ground. What is the speed of the ball when it leaves Sarah's hand? (f) How high above the ground will the ball be when it gets to Julie?

2. Soccer Kick: A soccer ball is kicked with an initial horizontal velocity of 18 m/s and an initial vertical velocity of 18 m/s. (a) What is the initial speed of the ball? (b) What is the initial angle θ of the ball with respect to the ground? (c) What is the maximum height the ball goes above the ground? (d) How far from where it was kicked will the ball land? (e) What is the speed of the ball 2.1 s after it was kicked? (f) How high above the ground is the ball 2.1 s after it is kicked?

FIGURE 2.11 Problem 2

3. Cannonball: A cannonball is shot (from ground level) with an initial horizontal velocity of 41 m/s and an initial vertical velocity of 27 m/s. (a) What is the initial speed of the cannonball? (b) What is the initial angle θ of the cannonball with respect to the ground? (c) What is the maximum height the cannonball goes above the ground? (d) How far from where it was shot will the cannonball land? (e) What is the speed of the cannonball 3.1 s after it was shot? (f) How high above the ground is the cannonball 3.1 s after it is shot?

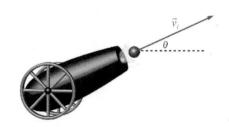

FIGURE 2.12 Problem 3

4. Baseball (INTERACTIVE EXAMPLE): Ted Williams hits a baseball with an initial velocity of 120 miles per hour (176 ft/s) at an angle of $\theta = 35°$ to the horizontal. The ball is struck 3 ft above home plate. You watch as the ball goes over the outfield wall 420 ft away and lands in the bleachers. After you congratulate Ted on his hit he tells you, "You think that was something? If there was no air resistance, I could have hit that ball clear out of the stadium!" Assuming Ted is correct, what is the maximum height of the stadium at its back wall (which is 565 ft from home plate) such that the ball would just pass over it? You may need: 9.8 m/s^2 = 32.2 ft/s^2 and 1 mile = 5,280 ft.

3

RELATIVE AND CIRCULAR MOTION

3.1 Overview

We will begin with a discussion of relative motion in one dimension. We will describe this motion in terms of displacement and velocity vectors which will allow us to generalize our results to two and three dimensions.

We will then discuss reference frames that are accelerating and will find that the description of phenomena in these frames can be somewhat confusing. For the remainder of the course, we will restrict ourselves to the description of phenomena in inertial reference frames, that is, those frames that are not accelerating.

Finally, we will consider a specific motion, namely uniform circular motion and find that it can be described in terms of a centripetal acceleration that depends on the angular velocity and the radius of the motion.

3.2   Relative Motion in One Dimension

In the last unit we discovered that two-dimensional projectile motion can be described as the superposition of two one-dimensional motions: namely, motion at constant acceleration in the vertical direction and motion at constant velocity in the horizontal direction. We then discussed the description of such a motion from two different reference frames. In particular, we saw that the motion of a ball thrown straight up on a train moving at constant velocity would be described as one-dimensional free fall by an observer on the train, but it would be described as a two-dimensional projectile motion by an observer on the ground. In this unit, we will develop a general equation that relates the description of a single motion in different reference frames.

We will start with a one-dimensional case. Suppose you are standing at a train station as a train passes by traveling east at a constant 30 m/s. Your friend Mike is on the train and is walking toward the back of the train at 1 m/s in the reference frame of the train. What is Mike's velocity in the reference frame of the station?

Intuitively, you probably realize the answer is 29 m/s, but it will prove useful to develop, in this one-dimensional example, a general procedure that can be easily generalized to the more non-intuitive cases that involve two or three dimensions.

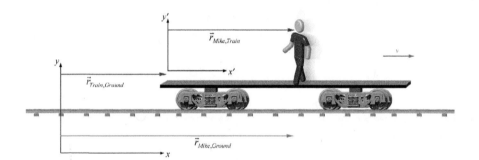

FIGURE 3.1 The displacement vectors of Mike in both the train and ground reference frames as well as the displacement vector of the origin of the train reference frame in the ground reference frame.

We start by drawing the displacement vectors of Mike in the two reference frames we have here: the train and the ground as shown in Figure 3.1. At any time the displacement of Mike with respect to the ground is just equal to the displacement of Mike with respect to the train plus the displacement of the train's origin with respect to the ground's origin.

$$\vec{r}_{Mike,Ground} = \vec{r}_{Mike,Train} + \vec{r}_{Train,Ground}$$

If we now differentiate this equation with respect to time, we obtain the equation we want: an equation that relates the velocities in the two frames. In particular, we see that the velocity of Mike with respect to the ground is equal to the velocity of Mike with respect to the train plus the velocity of the train with respect to the ground.

$$\vec{v}_{Mike,Ground} = \vec{v}_{Mike,Train} + \vec{v}_{Train,Ground}$$

Consequently, if we take east to be positive, then we see that Mike's velocity with respect to the ground is +29 m/s. Had Mike been walking toward the front of the train, his velocity would have been +31 m/s.

We hope this result seems reasonable and intuitive to you. Perhaps you're even wondering why we went to the trouble introducing a formalism to obtain a very intuitive result. We think the answer will become clear in the next section when we discuss relative motion in two and three dimensions.

3.3   Relative Motion in Two Dimensions

We now want to consider relative motion in more than one dimension. How do we go about solving this problem? Well, the formalism that we introduced in the last section is all we need to solve this problem. The key is that displacements, our starting point in the preceding section, are *vectors*! Therefore, when we differentiate the displacement equation, we obtained an equation that relates the velocities as vectors which holds in two and three dimensions as well!

Figure 3.2 shows Mike, in his motorboat, crossing a river. The river flows eastward at a speed of 2 m/s relative to the shore. Mike is heading north (relative to the water) and moves with a speed of 5 m/s (relative to the water). If you are standing on the shore, how do you describe his motion? How fast is he moving and in what direction?

To answer this question, we just apply our equation that relates the velocity vectors. Namely, Mike's velocity with respect to the shore is just the vector sum of his velocity with respect to the water (5 m/s due north) plus the velocity of the river with respect to the shore (2 m/s due east).

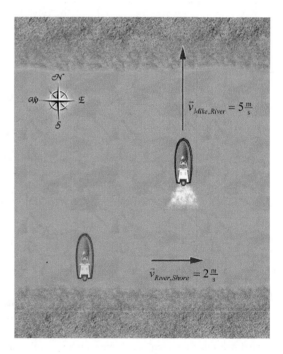

$$\vec{v}_{Mike,Shore} = \vec{v}_{Mike,River} + \vec{v}_{River,Shore}$$

Since these velocities are orthogonal, we can evaluate the vector sum using the

FIGURE 3.2 Mike heads due north moving at 5 m/s with respect to the river that itself flows due east at 2 m/s with respect to the shore.

Pythagorean theorem as shown in Figure 3.3. Namely, the magnitude of Mike's velocity with respect to the shore is just the length of the hypotenuse of the triangle. We can also obtain his direction from this right triangle by noting that $\tan\theta = 2/5$, where θ is the angle (east of north) that Mike moves as seen by an observer on the shore.

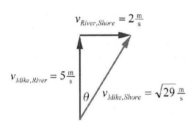

$$\theta = \tan^{-1}\frac{2}{5} = 21.8° \quad \text{(East of North)}$$

FIGURE 3.3 The determination of Mike's velocity with respect to the shore using vector addition.

This example illustrates all you have to know about relative motion. Writing down the displacement equation and differentiating it yields the relative velocity equation which can be applied in any relative motion problem.

As a note to the curious, though, this derivation does eventually break down once we consider speeds close to the speed of light. Namely, one of the postulates of special relativity states that the speed of light is the same in all reference frames. In order for the speed of light to be the same in all reference frames, time *cannot* be the same in all reference frames. Time is no longer a river; observers in reference frames that are in relative motion must report different measurements for the time intervals between the same two events. Consequently, our derivation of the simple relation between relative velocities will not work in special relativity. The expression that describes how velocities add in special relativity is given by

$$\vec{v}_{A,C} = \frac{(\vec{v}_{A,B} + \vec{v}_{B,C})}{1 + \left(\dfrac{v_{A,B}\,v_{B,C}}{c^2}\right)}$$

where c is the speed of light. Clearly, in the limit that the speeds $v_{A,B}$ and $v_{B,C}$ are small compared to that of light, this expression reduces to the form that we have just derived in the framework of Newtonian mechanics. Therefore, for our purposes in this course, we will simply use the Newtonian result for the addition of relative velocities.

$$\vec{v}_{A,C} = (\vec{v}_{A,B} + \vec{v}_{B,C})$$

3.4 Accelerating Reference Frames

In the last two sections we have related observations made in one reference frame with those in another. In both cases, one frame was moving relative to the other with a constant velocity. What would happen if this relative velocity was not constant? In other words, what happens in a reference frame which is accelerating?

We'll start with an everyday example. Suppose you have some fuzzy dice hanging from a string tied to the rearview mirror of your car. When you step on the gas and accelerate forward you see the dice swing backward as shown in Figure 3.4, just as if something were pushing them toward the back of the car. Indeed, you will feel yourself being pushed back into your seat in exactly the same way. Of course, no one is actually pushing the dice backward, nor is there anything actually pushing you back into the seat either. It just seems that way because you are accelerating.

FIGURE 3.4 As the car accelerates, the fuzzy dice swing backward.

Trying to understand physics in a reference frame that is accelerating can be confusing since the acceleration itself can easily be mistaken as a push or a pull. For this reason we usually only consider non-accelerating reference frames in this class, called **inertial reference frames**. We will develop this framework in the next unit. For now, though, we will take one more look at this example in order to discuss centripetal acceleration

3.5   Rotating Reference Frames

Suppose you are in the same car discussed in the last section. Imagine that instead of accelerating straight ahead, you are now driving with a constant speed, but you are going around a circular track turning to the left. What happens now to the dice suspended from your rearview mirror? Well, they will now swing over to the right! Since there is nothing actually pushing the dice, we conclude that the car, even though it is moving at a constant speed, must be accelerating.

We can visualize what is happening here if we consider a bird's eye view and draw in the velocity vector of the moving car as it moves as shown in Figure 3.5. The direction of this vector is clearly changing as a function of time, even though its magnitude, its speed, remains constant as a function of time. Since the acceleration vector is defined to be the time derivative of the velocity vector, we see that the acceleration of the car is non-zero! The velocity vector changes in time because its direction changes in time.

What is the direction and magnitude of the car's acceleration? By subtracting successive velocity vectors in the figure, we can see that the direction of this change in velocity is a vector that points toward the center of the circle. Since the direction of the acceleration is, by definition, the direction in which the velocity is changing, we see that the car's acceleration is toward the center of the circle. This result is consistent with the behavior of the dice hanging from the rearview mirror. Namely, just as before, they are tilted in the opposite direction of the acceleration.

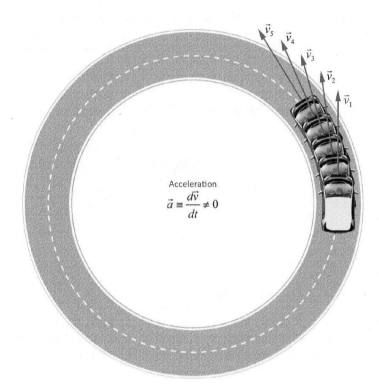

FIGURE 3.5 A car moves with constant speed around a circular track. Since the direction of the velocity vector is constantly changing, the acceleration of the car is not zero. In fact, the acceleration's magnitude is constant and its direction is always toward the center of the circle.

What about the magnitude of this acceleration? To determine the magnitude, we have to do some geometry. We'll start by considering two velocity vectors separated in time by a small amount equal to dt as shown in Figure 3.6. We define the angle between these velocity vectors during this time dt to be equal to $d\theta$. We can use the small angle approximation for $d\theta$ to obtain the magnitude of dv, the change in velocity vector during the time dt.

$$dv = v\, d\theta$$

The magnitude of the acceleration is just equal to the magnitude of the time rate of change of the velocity, namely, dv/dt. Therefore, we see that the magnitude of the acceleration is given by

$$\left|\vec{a}\right| \equiv \left|\frac{d\vec{v}}{dt}\right| = v\frac{d\theta}{dt}$$

We will discuss this acceleration and the meaning of $d\theta / dt$ in more detail in the next section.

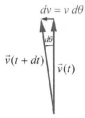

FIGURE 3.6 We use the small angle approximation to determine the magnitude of the centripetal acceleration.

3.6　Centripetal Acceleration

We just determined that an object following a circular path with a constant speed has an acceleration that points toward the center of the circle with a magnitude equal to the speed of the object times the time rate of change of its direction, $d\theta / dt$. Since the acceleration vector always points toward the center, we say this motion has **centripetal acceleration**.

It certainly makes sense that the acceleration the car experiences as it turns gets bigger as the speed increases, but what about the $d\theta / dt$ term? $d\theta / dt$, by definition, is equal to the time rate of change of θ, the angle that locates the car with respect to the center. Therefore, $d\theta / dt$ is a measure of how fast the direction of the car is changing in time, how fast it is turning. Put this way, it does make sense that the acceleration depends on $d\theta / dt$. Imagine driving at the same speed but turning the steering wheel harder and making your turn tighter; you are turning quicker so you expect the acceleration to feel bigger.

The time rate of change of θ, $d\theta / dt$, is called the **angular velocity** and is usually denoted by the symbol ω. This angular velocity can be determined from the speed v and the radius R that defines a particular uniform circular motion. Namely, the speed is just equal to the distance the car goes in one revolution, the circumference of the circle, divided by the time it takes to make one revolution, the **period**.

$$v = \frac{2\pi R}{T}$$

Now the angular velocity is equal to the total angular distance the car travels, 2π radians, divided by the time it takes to make one revolution, the period.

$$\omega = \frac{2\pi}{T}$$

Comparing these two equations, we see that the angular velocity is just equal to the velocity divided by the radius of the motion.

$$\omega = \frac{v}{R}$$

We can now write our expression for the centripetal acceleration in a more familiar way:

$$a = v\omega = \frac{v^2}{R}$$

3.7 Examples

We'll close this unit by doing a couple of examples to get a feeling for the magnitudes of accelerations that can be encountered.

We'll start by driving around a circular track whose radius is 100 m. How fast do we have to go in order for the magnitude of our centripetal acceleration to be one g, the acceleration due to gravity?

Setting our expression for the centripetal acceleration equal to 1 g (9.8 m/s^2), we find that the required velocity is equal to about 31 m/s or about 70 mph.

$$v = \sqrt{aR} = \sqrt{gR} = 31 \text{ m/s}$$

You might be interested to know that racecar drivers regularly experience even higher accelerations. For example, at the Indianapolis 500, typical accelerations in the turns are about 4 g's.

For our second example, we will consider the rotation of a person at rest on the surface of the Earth. This person, while at rest with respect to the surface of the Earth, is actually moving quite fast with respect to the Earth's axis. Indeed, this speed is easy to calculate: at the equator this speed is just equal to the circumference of the Earth divided by the time in one day, which works out to be around 1000 miles per hour!

$$v_{equator} = \frac{2\pi R_{Earth}}{T} = \frac{2\pi (6.4 \times 10^6 \text{ m})}{8.64 \times 10^4 \text{ s}} = 465 \text{ m/s}$$

This velocity is certainly significant, but the quantity of interest here is the centripetal acceleration. When we divide the square of this speed by the radius of the Earth, we find that the acceleration is, in fact, quite small: about one third of one percent of g. This value is small enough that we don't really notice it, and we are quite justified in ignoring its effect for any measurements we will make. We can and will assume that the surface of the Earth is a perfectly fine inertial reference frame.

By the way, we are also rotating about the Sun once per year. We leave it as an exercise for you to calculate the centripetal acceleration of this motion. You should find that it is even smaller than that of the rotation of the Earth about its axis.

MAIN POINTS

Relative Motion

The velocity of an object in frame A can be found from its velocity in frame B by adding (as vectors) the relative velocity of the two frames.

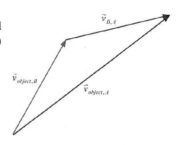

$$\vec{v}_{object,A} = \vec{v}_{object,B} + \vec{v}_{B,A}$$

Uniform Circular Motion

An object moving with constant speed in a circular path has an acceleration whose direction is always toward the center of the circle (centripetal acceleration) and whose magnitude is proportional to the square of the speed and inversely proportional to the radius of the circle.

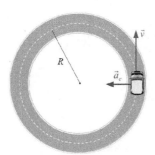

Centripetal Acceleration

$$a_c = \frac{v^2}{R} = \omega^2 R$$

PROBLEMS

1. Train Ride: You are on a train traveling east at a speed of 24 m/s with respect to the ground. (a) If you walk east toward the front of the train with a speed of 3.2 m/s with respect to the train, what is your velocity with respect to the ground? (b) If you walk west toward the back of the train with a speed of 3.8 m/s with respect to the train, what is your velocity with respect to the ground? (c) Your friend is sitting on another train traveling west at 18 m/s. As you walk toward the back of your train at 3.8 m/s, what is your velocity with respect to your friend?

2. Plane Ride: You are on an airplane traveling due east at 120 m/s with respect to the air. The air is moving with a speed of 37 m/s with respect to the ground at an angle of 30° west of due north. (a) What is the speed of the plane with respect to the ground? (b) What is the heading of the plane with respect to the ground? (Let 0° represent due north, 90° represent due east). (c) How far east will the plane travel in 1 hour?

3. Car on Curve: A car is traveling around a horizontal circular track with radius $r = 270$ m at a constant speed of $v = 29$ m/s as shown. The angle $\theta_A = 15°$ is above the x-axis, and the angle $\theta_B = 54°$ is below the x-axis. (a) What is the magnitude of the car's acceleration? (b) What is the x component of the car's acceleration when it is at point A? (c) What is the y component of the car's acceleration when it is at point A? (d) What is the x component of the car's acceleration when it is at point B? (e) What is the y component of the car's acceleration when it is at point B? (f) As the car passes point B, is the y component of its acceleration *increasing*, *constant*, or *decreasing*?

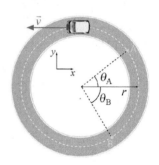

FIGURE 3.7 Problem 3

4. River Rescue (INTERACTIVE EXAMPLE): A child is in danger of drowning in the Merimac River. The Merimac River has a current of 3.1 km/hr to the east. The child is 0.6 km from the shore and 2.5 km upstream from the dock. A rescue boat with a speed of 24.8 km/hr (with respect to the water) sets off from the dock at the optimum angle to reach the child as fast as possible. How far from the dock does the boat reach the child?

FIGURE 3.8 Problem 4

4

NEWTON'S LAWS

4.1 Overview

The last unit marked the completion of our study of kinematics, the description of motions in terms of displacements, velocities and accelerations. We will now begin our study of dynamics, an account of what causes these motions.

The framework we will use was first introduced by Isaac Newton in 1687 in his major work, *Principia Mathematica*. We will use what are commonly called Newton's three laws of motion to develop this framework. We will begin with Newton's second law which relates two new physics concepts, *force* and *mass*, that are needed to determine the motion of an object in any given physical situation. We will then recast Newton's second law in terms of another important physics concept, momentum.

We will then introduce Newton's first law which establishes the concept of inertial reference frames. We will conclude by introducing Newton's third law, which states that all forces come in pairs: that for every action, there is an equal and opposite reaction.

4.2 Two Concepts

Up to this point our focus has been *kinematics*, the description of motion. We now want to shift our focus to **dynamics**, the causes of motion. In order to discuss the causes of motion, Newton introduced two new concepts that were not needed in our study of kinematics.

These two new concepts are called *mass* and *force*. Although these words are in common use in the English language, in physics, these words have very specific meanings. We will need to be very careful when we use these words to make sure that we are referring to their physics meanings, which are much more restricted than their ordinary language meanings.

We will start with **mass**. *Mass is the property of an object that determines how hard it is to change its velocity.* Objects that are difficult to accelerate have large mass, while objects that are easy to accelerate have small mass.

The second concept is **force**. *Force is the thing that is responsible for an object's change in velocity.* It's common to think of forces in terms of pushes and pulls, but we will find that we must extend our definition of force beyond that of the exertion made by people to move objects. We will need to think of forces simply in terms of producing a change in an object's velocity.

The most important thing to notice in these definitions is that *both* mass and force refer to changing an object's velocity, that is, to its *acceleration*. Acceleration is the kinematic concept that links these two dynamic concepts. This link is formalized in Newton's second law which we will now introduce.

4.3 Newton's Second Law

Newton's insight, which is captured in his second law is that when a force acts on an object it causes that object to accelerate in the same direction that the force acts, and that the magnitude of this acceleration is proportional to the magnitude of the force.

$$\vec{a} = \frac{\vec{F}}{m}$$

Note that the constant of proportionality in this equation is 1/mass. It is the mass of an object that determines the acceleration for a given applied force. The bigger the mass, the smaller the acceleration for a given force.

We have introduced the form of Newton's second law as $\vec{a} = \vec{F}/m$ rather than the more familiar (and equivalent) $\vec{F} = m\vec{a}$, in order to stress that it is the force that causes the acceleration, and not vice versa. Often students want to invoke an "$m\vec{a}$" force in their problem solutions. By writing $\vec{a} = \vec{F}/m$, we are encouraging you to think of the forces as being primary and the acceleration as the result of applying the forces to the mass.

Finally, we want to stress that this equation is a *vector* equation. The importance of the vector nature of this equation can be demonstrated by two observations. First, the direction of the acceleration is the same as the direction of the force. Second, we exploit the vector nature of forces to determine what happens when more than one force acts on an object. Namely, if more than one force acts on an object, the object's acceleration is determined by applying Newton's second law using the total force acting on the object which is defined to be the vector sum of all the individual forces acting on the object.

$$\vec{F}_{Net} = \sum_i \vec{F}_i$$

4.4 Units

We have just introduced the two new concepts we need to describe dynamics (namely, force and mass) and the fundamental law (Newton's second law) that links them. We now need to address the issue of units.

So far, we have used two basic units, one for space (meters) and one for time (seconds) to describe kinematics. To accommodate the two new concepts needed to describe dynamics, we will need to introduce a new unit. We choose this new unit, the *kilogram*, to be the SI unit for mass. Using Newton's second law, we see that the SI unit for force is $kg \cdot m/s^2$. It is common to define the newton (N), a unit of force, to be equal to $1\ kg \cdot m/s^2$.

The truly amazing thing here is that during this course in mechanics, we will introduce many new concepts, but we will need to introduce *no more fundamental units*! Indeed, we need just two kinematics units (meters and seconds) and one dynamic unit (mass) to specify completely all of the concepts we will introduce in mechanics. Next term, when we study electricity and magnetism, we will need to introduce only one more unit, the coulomb, the unit for electric charge.

4.5 Momentum

Before moving on to do an example that illustrates Newton's second law, we first want to make a natural connection between forces and another physics quantity of fundamental importance: *momentum*.

Recalling that the acceleration is defined to be the time derivative of the velocity, we can rewrite Newton's second law in terms of this derivative. Now, as long as the mass is a constant, we can bring it inside the derivative to obtain the expression shown.

$$\vec{F}_{Net} = m\vec{a} = m\frac{d\vec{v}}{dt} = \frac{d}{dt}(m\vec{v}) = \frac{d\vec{p}}{dt}$$

The expression inside the parentheses (the product of the object's mass and its velocity) is defined to be the **momentum** of the object, usually denoted by the symbol $\vec{p}$. With this definition, we see that Newton's second law is equivalent to the statement that the time rate of change of an object's momentum is equal to the total force acting on the object. In fact, when Newton first introduced this law in his *Principia* in 1687, he used this formulation in terms of the change in momentum. One advantage of this formulation is that it can be used for processes in which the mass of the system is changing in time.

There are two simple, yet very interesting conclusions we can draw from this formulation of Newton's second law in terms of the change in momentum. First, if there are no forces acting on an object then its momentum cannot change. Indeed, since Newton's law is a vector equation, this statement must be individually true for all components of the momentum. Second, to change an object's momentum, we need to have a non-zero force acting on the object for a finite period of time. We will revisit both of these concepts later in this course when we discuss collisions.

4.6   Example: Spaceship

We will now return to the usual formulation of Newton's second law to do a simple example that illustrates some of its main features. Consider a spaceship having a constant mass m far out in space. Suppose the engines on the spaceship are turned on at $t = 0$ and are then turned off 50 seconds later, and that during the time the engines are on they exert a force $\vec{F}$ on the spaceship. Figure 4.1 shows the plots of the displacement, velocity and acceleration of the spaceship as a function of time.

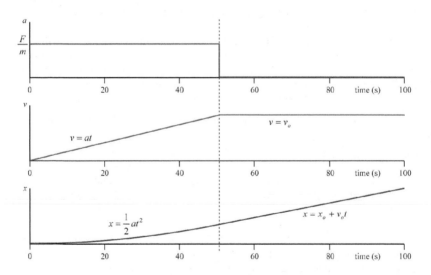

FIGURE 4.1 Plots of the acceleration (purple), velocity (blue), and displacement (red) of a spaceship of mass m that experiences a constant net force $\vec{F}$ for a fixed time (50 s).

While the engines are on, the spaceship will have a constant acceleration equal to $\vec{F}/m$. The acceleration is just equal to $\vec{F}/m$ between 0 and 50 seconds, and zero thereafter.

While the spaceship has a constant non-zero acceleration, the velocity increases linearly with time as shown. During this time the displacement increases quadratically as shown.

In fact, these plots are just the plots of motion at constant acceleration that you are familiar with from kinematics. The only new thing here is that we *know* that the acceleration is constant, because the force (and mass) is constant.

After the engines shut off at $t = 50$ s, the acceleration goes to zero. Therefore, the motion for $t > 50$ s is one of constant velocity. Consequently, the displacement, for $t > 50$ s, just increases linearly.

4.7 Newton's First Law

Newton's second law is, in some sense, the "working equation" of this course. You will use it time and time again to solve problems. In his famous book, the *Principia*, Newton introduced two other laws, though, that we want to discuss at this time.

Newton's first law states that "a body at rest or in uniform motion in a straight line will continue that motion unless acted upon by an unbalanced force." We have, in fact, already used this formulation to define force as "the thing that is responsible for an object's change in velocity." Consequently, in this course, we will use a formulation of Newton's first law that explicitly introduces a new and important concept: an *inertial reference frame*. In particular, we will say "an object subject to no external forces is at rest or moves with constant velocity if viewed from an inertial reference frame."

This formulation of Newton's first law explicitly refers to the reference frame in which the motion is described. In particular, Newton's first law actually serves to define an inertial reference frame as a reference frame in which Newton's laws hold good. We know Newton's laws do not appear to hold in all reference frames. For example, we have noted that objects appear to behave strangely in accelerating frames, as though forces were acting on them, even though we could not identify the agent that produced these forces. If you turn a tight corner in your car, the cell phone on your dashboard may slide from one side to the other, clearly accelerating, even though there are not any sudden new forces acting on it that we could identify to be the cause of this acceleration.

For our current purposes, we can identify inertial frames as non-accelerating frames. We want to relax this identification a bit, though, as we have already seen that the Earth's acceleration as it spins on its axis is not zero, but it is sufficiently small that we may, for practical purposes, consider the Earth to be an inertial reference frame.

We will close this discussion by noting that Newton's First Law implies that once we have found one inertial reference frame, then any other reference frame moving with a constant velocity with respect to the first one is also an inertial reference frame. Therefore, accepting the Earth as an inertial reference frame implies that any frame moving at constant velocity with respect to the Earth is also an inertial reference frame.

4.8 Newton's Third Law

Newton's third law states that "for every action there is an equal and opposite reaction." This sentence can be misleading, but its meaning is succinctly and completely captured in the vector equation shown.

$$\vec{F}_{AB} = -\vec{F}_{BA}$$

Namely, that the force object A exerts on object B is equal to minus the force object B exerts on object A.

The important point to realize here is that Newton's third law implies that all forces come in pairs! There is no such thing as an isolated force. If you exert a force on the wall by pushing on it, the wall exerts a force on you by pushing back with a force of exactly the same magnitude, but in the exact opposite direction. Figure 4.2 shows the forces that are exerted in this situation. Our notation for force labels uses subscripts to denote what is causing the force and what the force is pushing against. $\vec{F}_{Man,Wall}$ therefore refers to the force exerted *by* the man *on* the wall, and $\vec{F}_{Floor,Man}$ refers to the force exerted *by* the floor *on* the man. With this notation, it is clear which forces are the third-law "pairs." For every $\vec{F}_{AB}$ there must be an equal and opposite $\vec{F}_{BA}$.

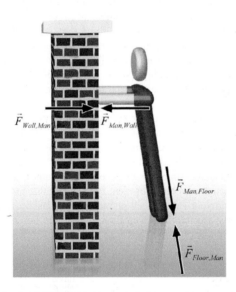

FIGURE 4.2 Newton's third law pairs of forces acting when a man stands on the floor and pushes against the wall.

Consider the case of a man pushing on a box initially at rest on a smooth horizontal surface, causing it to accelerate to the left as shown in Figure 4.3. The man pushes on the box with a force $\vec{F}_{mb}$ and the box pushes back on the man with an equal and opposite force $\vec{F}_{bm}$. When we add together all of the horizontal forces they seem to cancel. But if that were true, the box would not accelerate. But we know it will! What's going on here? Well, the answer to our problem is simple. In order to understand the motion of the *box* we need to consider only the forces acting *on* the *box*–only $\vec{F}_{mb}$ needs to be considered when calculating the total force on the box, not $\vec{F}_{bm}$. This little exercise may seem more like bookkeeping than physics, but it is extremely important to get this picture correct at the outset. All forces come in pairs, but these force pairs are exerted on *different objects*.

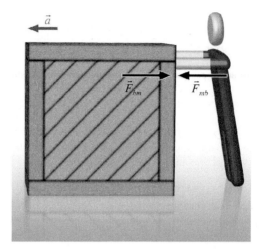

FIGURE 4.3 A man exerts a horizontal force $\vec{F}_{mb}$ on a box, causing it to accelerate. The force $\vec{F}_{bm}$ acts on the man, not the box; therefore, it plays no role in the determination of the acceleration of the box.

To determine the motion of an object, we only need to consider the forces acting on that object.

In the next unit we will develop the technique of free-body diagrams that will formalize this notion and lay the groundwork for all problem solving using Newton's second law.

MAIN POINTS

Force, Mass and Newton's Second Law

Mass is the property of an object that determines how hard it is to change its velocity.

Force is the thing that is responsible for an object's change in velocity.

Newton's second law: A force that acts on an object causes that object to accelerate in the same direction that the force acts, and the magnitude of this acceleration is proportional to the magnitude of the force

$$\vec{a} = \frac{\vec{F}_{Net}}{m}$$

This law provides the link between mass and force.

Inertial Reference Frames and Newton's First Law

Newton's first law: An object subject to no external forces is at rest or moves with constant velocity if viewed from an inertial reference frame. This law serves to define "inertial reference frames."

Newton's Third Law

Newton's third law: For every action there is an equal and opposite reaction.

All forces come in pairs but act on different objects.

$$\vec{F}_{AB} = -\vec{F}_{BA}$$

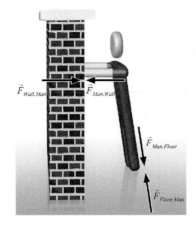

PROBLEMS

1. Take Off: A 120,000 kg jet accelerates down the runway for takeoff at 1.6 m/s^2. (a) What is the net horizontal force on the airplane as it accelerates for takeoff? (b) What is the net vertical force on the airplane as it accelerates for takeoff? (c) Once off the ground, the plane climbs upward for 20 s. During this time, the vertical speed increases from zero to 25 m/s, while the horizontal speed increases from 80 m/s to 91 m/s. What is the net horizontal force on the airplane as it climbs upward? (d) What is the net vertical force on the airplane as it climbs upward? (e) After reaching cruising altitude, the plane levels off, keeping the horizontal speed constant but smoothly reducing the vertical speed to zero, in 11 s. What is the net horizontal force on the airplane as it levels off? (f) What is the net vertical force on the airplane as it levels off?

2. Three Blocks: Three blocks, each of mass 13 kg are on a frictionless table. A hand pushes on the left most block (block A) such that the three boxes accelerate in the positive horizontal direction as shown at a rate of $a = 0.7$ m/s^2.

FIGURE 4.4 Problem 2

(a) What is the magnitude of the force on block A from the hand? (b) What is the net horizontal force on block A? (c) What is the horizontal force on block A due to block B? (d) What is the net horizontal force on block B? (e) What is the horizontal force on block B due to block C?

3. Three Boxes: Three boxes, each of mass 14 kg are on a frictionless table, connected by massless strings. A force of tension T_1 pulls on the right most box (box A) such that the three boxes accelerate in the positive horizontal direction at a rate of $a = 0.7$ m/s^2.

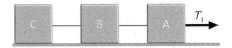

FIGURE 4.5 Problem 3

(a) What is the magnitude of T_1? (b) What is the net horizontal force on box A? (c) What is the net force that box B exerts on box A? (d) What is the net horizontal force on box B? (e) What is the net force box C exerts on box B?

5

FORCES AND FREE-BODY DIAGRAMS

5.1 Overview

We will begin by introducing the bulk of the new forces we will use in this course. We will start with the weight of an object, the gravitational force near the surface of the Earth, and then move on to discuss the normal force, the force perpendicular to the surface that two objects in contact exert on each other, and the tension force, the force exerted by a taut string. Finally, we will introduce Newton's universal law of gravitation that describes the forces between any two objects that have mass. We will close by introducing free-body diagrams which we will then use in the solution of a Newton's second law problem.

5.2 Weight

In order to apply Newton's second law in physical situations, we will need to increase our inventory of forces. We will start with the gravitational force near the surface of the Earth.

We have already seen that an object in free fall near the surface of the Earth has a constant acceleration whose direction is down and whose magnitude is equal to the constant g, which is equal to about 9.8 m/s^2. From this description of the motion, we can use Newton's second law to conclude that there must be a force in the downward direction acting on the object and that the magnitude of this force must be equal to the product of the mass of the object and the constant g.

$$\vec{W} = m\vec{g}$$

We call this force the **weight** of the object. It is important to realize the weight of an object is NOT the same thing as its mass! Mass is an intrinsic property of the object; its value determines how hard it is to change its velocity. Mass does NOT depend on the location of that object or on its surroundings. Weight, on the other hand, just tells us the magnitude of the gravitational force that is acting on the object. We will investigate the nature of this gravitational force more fully after we first introduce a few more straightforward forces.

5.3 Support Forces: The Normal Force and Tension

We can use our knowledge of the weight force from the last section to motivate the need for two more forces. First, consider the incredibly mundane situation of a heavy box sitting on a floor as shown in Figure 5.1. What forces are acting on this box? Well, certainly the weight of the box is acting, supplying a force vertically downward. This cannot be the only force on the box, though, since if it were, Newton's second law would tell us that the box should be accelerating downward with constant acceleration equal to g. Therefore, to obtain the needed zero acceleration, there must be another force that acts vertically upward with the same magnitude. Note that this force is NOT the Newton's third law pair to the weight since both forces act on the same object, the box. This force is the force exerted by the floor on the box and is usually called the **normal force** since its direction is perpendicular to the surface.

What determines the magnitude of this force, in general? Well, to determine the magnitude of the normal force in any particular case, we do just as we did here; we apply Newton's law. *The normal force is simply what is has to be to do what it does*, which is to supply a supporting force for objects!

The total force exerted by any surface in contact with another surface will always have a normal component, but it

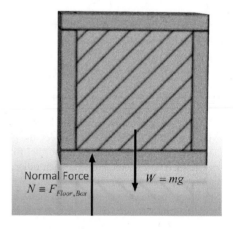

FIGURE 5.1 Two forces act on a box that is at rest on the floor: the weight W which is the gravitational force exerted by the Earth, and the normal force N exerted by the floor.

may also have a component parallel to the surfaces called the frictional force. We will discuss the nature of frictional forces in the next unit.

A force similar to the normal force between surfaces in contact with each other is the tension force in strings, wires and ropes. Figure 5.2 shows a ball hanging from a string. What forces are acting on the ball? Clearly the weight force acts vertically downward. In order to obtain the needed zero acceleration, the string must also be exerting a force on the ball. We call this force the **tension force**; it exists whenever the string is taut and its direction is along the string, in this case, vertically upward. From Newton's second law, we can determine that the magnitude of this force must be equal to the weight of the ball in order to provide the observed zero acceleration. Just as was the case for the normal force, *the tension force is simply what is has to be to do what it does!* In this case, the string is just holding up the ball; strings can also be used to pull objects across a surface. In either case, the magnitude of the tension must be determined from Newton's second law.

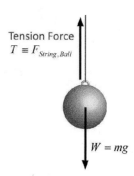

FIGURE 5.2 Two forces act on a ball that is suspended by a string: the weight W which is the gravitational force exerted by the Earth and the tension force T exerted by the string.

5.4   Springs

Figure 5.3 shows a ball hanging from a spring. We can use Newton's second law to determine that the spring must be exerting a force on the ball that is equal to its weight. However, if we were to replace the first ball with a new ball that has twice the mass, we would see that the spring would be stretched more. In this new situation, we know the spring would be exerting twice the force, but its length would also be increased. In fact, the amount by which the length changes tells us the magnitude of the force! The key concept here is that every spring has an equilibrium length, and if it is stretched or compressed by some amount Δx from this length, it will exert a restoring force that opposes this change. The magnitude of this force is proportional to Δx, the extension or compression of the spring from its equilibrium position.

The force law for springs that quantifies this relation is given by

$$\vec{F}_{spring} = -k(\vec{x} - \vec{x}_o)$$

where the vector $\vec{x}_o$ represents the equilibrium length of the spring, while the vector $\vec{x}$ represents the final length of the spring. The vector difference $\vec{x} - \vec{x}_o$ represents the amount by which the spring is either stretched or compressed. The minus sign in the equation illustrates that the force is always in the opposite direction of the vector $\vec{x} - \vec{x}_o$, the extension or compression of the spring. If the spring is stretched, $\vec{x} - \vec{x}_o$ points away from the equilibrium position; therefore, the force is directed back toward the equilibrium

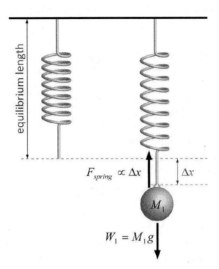

FIGURE 5.3 Two forces act on a ball that is suspended by a spring: the weight W, and the force exerted by the spring F_{spring}. The magnitude of the spring force is proportional to the extension (or compression) from the spring's equilibrium position.

position. If the spring is compressed, $\vec{x} - \vec{x}_o$ once again points away from the equilibrium position; therefore, the force is again directed back toward the equilibrium position. Since the force exerted by the spring is always directed toward its equilibrium position, we call this force a *restoring force*. The directions of these vectors are illustrated in Figure 5.4.

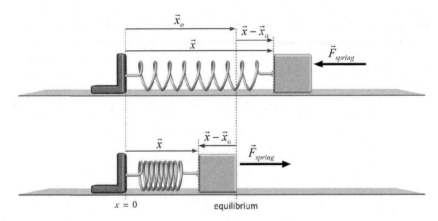

FIGURE 5.4 The spring force is a restoring force. The force vector ($\vec{F}_{spring} = -k(\vec{x} - \vec{x}_o)$) always points back toward the equilibrium position as illustrated in extension (top) or compression (bottom).

If we define the origin of our coordinate system to be at the equilibrium position, then our force equation simplifies to

$$\vec{F} = -k\vec{x}$$

The symbol k stands for the **spring constant** of the particular spring and is a measure of its stiffness. The units of k are newtons per meter (N/m); a large value of k means that a small deformation results in a big force.

5.5   Universal Gravitation

We now want to generalize our earlier discussion of the force we called weight. You know that the Moon orbits the Earth with a period of about a month, and the Earth orbits the Sun with a period of a year. To a good approximation these orbits are examples of uniform circular motion, and therefore, we know that each orbiting body experiences a centripetal acceleration. In Newton's framework, there must be a real force being exerted on the orbiting body that is responsible for this acceleration. Newton proposed that this force was a **universal gravitational force** that exists between any two objects that have mass.

In particular, he said that any two objects with mass exert attractive forces on each other with a magnitude that is proportional to the product of the masses divided by the square of the distance between them and whose direction lies along a line connecting them (see Figure 5.5).

$$\vec{F}_{12} = -G\frac{m_1 m_2}{r_{12}^2}\hat{r}_{12}$$

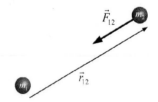

FIGURE 5.5 Mass m_1 (blue) exerts a gravitational force $\vec{F}_{12}$ on mass m_2 (red) that is proportional to the product of both masses and inversely proportional to the square of the distance between them. The gravitational force is an attractive force.

In this expression, $\vec{F}_{12}$ is the gravitational force from m_1 on m_2, G represents the **universal gravitational constant**, and $\hat{r}_{12}$ is the unit vector in the direction from m_1 to m_2.

Figure 5.6 illustrates the application of this expression to the Earth-Moon system. The symbols M_E and M_m refer to the masses of the Earth and Moon respectively, while R_{Em} is the Earth-Moon distance. We know the magnitude of the acceleration of the Moon, a_m, is equal to the square of its speed divided by the Earth-Moon distance. Applying Newton's second law, we can determine the acceleration of the Moon.

$$a_m = \frac{v_m^2}{R_{Em}} = \frac{F_{Em}}{M_m} = G\frac{M_E}{R_{Em}^2}$$

All quantities in this expression were known to Newton except the universal gravitational constant and the mass of the Earth.

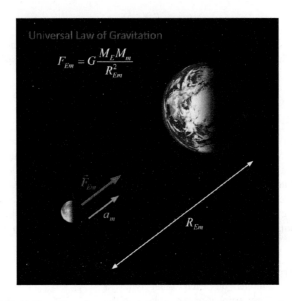

FIGURE 5.6 The universal gravitational force exerted by the Earth on the Moon provides the necessary centripetal acceleration to keep the Moon in its orbit about the Earth.

Newton, however, realized that the known acceleration due to gravity near the surface of the Earth was also proportional to the product of these unknown quantities! In order to make this realization, though, he essentially had to invent the calculus to show that the force the Earth exerts on any object is equivalent to that obtained by simply placing all of the mass of the Earth at its center.

$$Weight = mg = G\frac{M_E m}{R_E^2}$$

Given this result, we see that the acceleration due to gravity near the surface of the Earth is equal to the product of the universal gravitational constant and the mass of the Earth divided by the square of the radius of the Earth.

$$g \equiv G\frac{M_E}{R_E^2}$$

Therefore, we see that the ratio of the Moon's acceleration to that of an apple in free fall near the surface of the Earth is predicted to be equal to the ratio of squares of the radius of the Earth to the Earth-Moon distance.

$$\frac{a_m}{g} = \frac{R_E^2}{R_{Em}^2}$$

Now, the speed of the Moon in its orbit is 1.02 km/s, while the Earth-Moon distance is 3.844×10^5 km and the radius of the Earth is 6,371 km. When we plug in these numbers for the known quantities, we find that this prediction is verified! This result is really amazing! It represents the first demonstration that the same physical laws that operate here on Earth also operate in the heavens!

5.6 Free-Body Diagrams

In the last unit, we introduced Newton's three laws which supply the framework we will use to develop our understanding of dynamics. In particular, these laws will provide the basis for our understanding of the motion of any object in terms of the forces that act on it. In order to use these laws successfully, though, we need to keep careful track of the magnitudes and the directions of all forces acting on the object in question; we will use free-body diagrams to accomplish this task.

Figure 5.7 shows a man pushing a box across a smooth floor with a representation of all forces that are acting. Contact forces are shown in red, and the gravitational forces are shown in blue. Note that all of forces come in pairs, as required by Newton's third law. For example, the force exerted by the box on the man is equal and opposite to the force exerted by the man on the box.

We would like to calculate the acceleration of the box. How do we go about making this calculation? The key step here is to realize that the *only* forces which are relevant to this

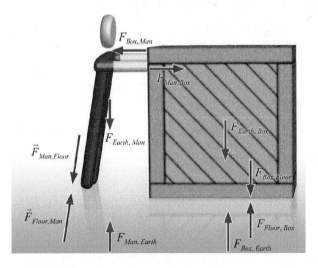

FIGURE 5.7 All the forces that act when a man pushes a box across a smooth floor.

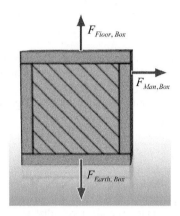

FIGURE 5.8 The free-body diagram for the box shown in Figure 5.7.

problem are the ones that act *on* the box–all other forces can be ignored. A diagram showing only these forces is called a **free-body diagram** for the box and is illustrated in Figure 5.8.

Applying Newton's second law to the box, we see that the acceleration of the box will be equal to the total force on the box divided by the mass of the box. To determine the total force on the box, we will need to add, as vectors, all of the forces shown in this free-body diagram for the box. Usually, in order to add these vectors, we will want to decompose the forces into appropriate components and write Newton's second law equations for each component separately.

$$a_{horizontal} = \frac{F_{Net,horizontal}}{M_{Box}} \qquad \rightarrow \qquad a_{horizontal} = \frac{F_{Man,Box}}{M_{Box}}$$

$$a_{vertical} = \frac{F_{Net,vertical}}{M_{Box}} \qquad \rightarrow \qquad 0 = \frac{F_{Floor,Box} - F_{Earth,Box}}{M_{Box}}$$

5.7   Example: Accelerating Elevator

We will close this unit by considering a one-dimensional problem to illustrate the procedure to use when solving dynamics problems.

Figure 5.9 shows a box of mass m hanging by a rope from the ceiling of an elevator moving vertically with acceleration a. We want to calculate the tension in the rope for any value of this acceleration.

The first step is to draw a picture and label all the forces acting on the object in question. In this case the object is the box, and the forces acting on it are the tension in the rope (T), which points upward, and the weight of the box (mg), which points downward, as shown

The second step is to choose a coordinate system. Any system will do, but you will soon discover that choosing one in which one of the axes is parallel to the acceleration will simplify the calculation.

The next step is to use your picture as a guide to write down the components of Newton's second law and solve for whatever variable you want to determine. In our case all of the forces act along a single direction so that we only have one equation to solve. The force

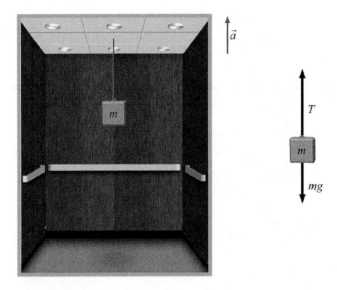

FIGURE 5.9 A box of mass m hangs by a rope in an elevator that is moving vertically with acceleration $\vec{a}$. To find the tension in the rope, we use the free-body diagram for the box that is shown.

on the box due to the rope is T in the $+y$ direction, and the force on the box due to gravity is mg in the $-y$ direction; therefore the total force on the box in the $+y$ direction is given by

$$F_{Net,y} = T - mg$$

Substituting this expression for the total force in Newton's second law yields the result that the tension is equal to the weight of the box plus the product of the mass of the box and its acceleration.

$$T = m(a + g)$$

The final step is to check to see if your answer makes sense. In this case we just found that the tension in the rope is given by the weight of the box plus an extra part which is proportional to the acceleration. If we consider the case where the elevator is not accelerating, we see that the tension in the rope is just equal to the weight of the box. If the elevator is accelerating upward, the tension is bigger than the weight, and if the elevator is accelerating downward the tension is less than the weight. All of these observations make sense.

MAIN POINTS

Support Forces

Support forces, for example the normal force and the tension force, are what they have to be to do what they have to do. Magnitudes are determined by Newton's second law.

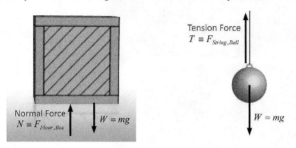

Spring Force

The spring force is a restoring force.

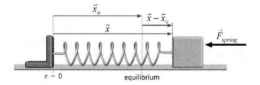

$$\vec{F}_{spring} = -k\left(\vec{x} - \vec{x}_o\right)$$

Universal Gravitation

Any two objects with mass exert attractive forces on each other with a magnitude that is proportional to the product of the masses divided by the square of the distance between them and whose direction lies along a line connecting them.

$$F_{gravity} = G\frac{m_1 m_2}{r^2}$$

Acceleration due to gravity (near the Earth's surface):

$$g \equiv G\frac{M_{Earth}}{R^2_{Earth}} = 9.8\,\tfrac{m}{s^2}$$

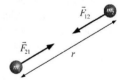

PROBLEMS

1. Hanging Masses: A single mass $m_1 = 4.2$ kg hangs from a spring in a motionless elevator. The spring is extended $x = 16$ cm from its unstretched length. (a) What is the spring constant of the spring? (b) Now, three masses $m_1 = 4.2$ kg, $m_2 = 12.6$ kg, and $m_3 = 8.4$ kg hang from three identical springs in a motionless elevator. The springs all have the same spring constant that was calculated in part (a). What is the force the top spring exerts on the top mass? (c) What is the distance the lower spring is stretched from its equilibrium length? (d) Now the elevator is moving downward with a velocity of $v = -3.1$ m/s but accelerating upward at an acceleration of $a = 3.8$ m/s^2. (*Note*: An upward acceleration when the elevator is moving down means the elevator is slowing down.) What is the force the bottom spring exerts on the bottom mass? (e) What is the distance the upper spring is extended from its

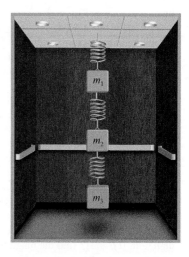

FIGURE 5.10 Problem 1 parts (b)-(h)

unstretched length? (f) Finally, the elevator is moving downward with a velocity of $v = -3.4$ m/s and also accelerating downward at an acceleration of $a = -2.0$ m/s^2. Is the elevator *speeding up*, *slowing down*, or *moving at a constant speed*? (g) Rank the distances the springs are extended from their unstretched lengths: $x_1 = x_2 = x_3$, $x_1 > x_2 > x_3$, or $x_1 < x_2 < x_3$. (h) What is the distance the *middle* spring is extended from its unstretched length?

2. Loop-the-Loop: In a loop-the-loop ride, a car goes around a vertical circular loop at a constant speed. The car has a mass $m = 268$ kg and moves with a speed of $v = 16.08$ m/s. The loop-the-loop has a radius of $R = 10.3$ m. (a) What is the magnitude of the normal force on the car when it is at the bottom of the circle and is accelerating upward? (b) What is the magnitude of the normal force on the car when it is at the side of the circle moving vertically upward? (c) What is the magnitude of the normal force on the car when it is at the top of the circle? (d) Compare the magnitude of the car's acceleration at each of the above locations:

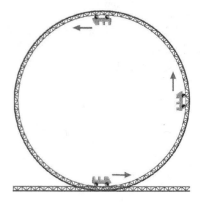

FIGURE 5.11 Problem 2

$a_{bottom} = a_{side} = a_{top}$, $a_{bottom} < a_{side} < a_{top}$, or $a_{bottom} > a_{side} > a_{top}$. (e) What is the minimum speed of the car so that it stays in contact with the track at the top of the loop?

3. Tennis Ball: A tennis ball connected to a string is spun around in a vertical, circular path at a uniform speed. The ball has a mass $m = 0.153$ kg and moves at $v = 4.89$ m/s. The circular path has a radius of $R = 0.94$ m. (a) What is the magnitude of the tension in the string when the ball is at the bottom of the circle? (b) What is the magnitude of the tension in the string when the ball is at the side of the circle? (c) What is the magnitude of the tension in the string when the ball is at the top of the circle? (d) Compare the magnitude of the ball's acceleration at each of the above locations: $a_{bottom} = a_{side} = a_{top}$, $a_{bottom} < a_{side} < a_{top}$, or $a_{bottom} > a_{side} > a_{top}$. (e) What is the minimum velocity so the string will not go slack as the ball moves around the circle?

4. Two Masses Over Pulley: A mass $m_1 = 6.2$ kg rests on a frictionless table. It is connected by a massless and frictionless pulley to a second mass $m_2 = 2.3$ kg that hangs freely. (a) What is the magnitude of the acceleration of block 1? (b) What is the tension in the string? (c) Now the table is tilted at an angle of $\theta = 80°$ with respect to the vertical. Find the magnitude of the new acceleration of block 1. (d) At what "critical" angle will the blocks NOT accelerate at all? (e) Now the angle is decreased past the "critical" angle so the system accelerates in the opposite direction. If $\theta = 31°$ find the magnitude of the acceleration. (f) Compare the tension in the string in each of the above cases on the incline: $T_{\theta \text{ at } 80°} = T_{\theta_{critical}} = T_{\theta \text{ at } 31°}$, $T_{\theta \text{ at } 80°} > T_{\theta_{critical}} > T_{\theta \text{ at } 31°}$, or $T_{\theta \text{ at } 80°} < T_{\theta_{critical}} < T_{\theta \text{ at } 31°}$.

FIGURE 5.12 Problem 4 parts (c)-(f)

5. Satellite in Orbit: Scientists want to place a 2,800 kg satellite in orbit around Mars. They plan to have the satellite orbit a distance equal to 2.4 times the radius of Mars above the surface of the planet. Here is some information that will help solve this problem: $m_{Mars} = 6.4191 \times 10^{23}$ kg, $r_{Mars} = 3.397 \times 10^6$ m, and $G = 6.67428 \times 10^{-11}$ N·m²/kg². (a) What is the force of attraction between Mars and the satellite? (b) What speed should the satellite have to be in a perfectly circular orbit? (c) How much time does it take the satellite to complete one revolution? (d) Which of the following quantities would change the speed the satellite needs to orbit: *the mass of the satellite, the mass of the planet,* and/or *the radius of the orbit*? (e) What should the radius of the orbit be (measured from the center of Mars), if we want the satellite to take eight times longer to complete one full revolution of its orbit?

UNIT

6

FRICTION

6.1 Overview

This unit is devoted to the introduction of a single force, friction. Frictional forces are omnipresent in our world. Indeed, pretty much all science prior to Galileo, focused on what was directly observed, much of that being dominated by friction. Galileo, in his description of free fall, and Newton in his first law, took as fundamental the more idealized description of motions in the *absence* of friction. The effects of friction then can be "added back in." We will learn in this unit exactly how to account for these effects of friction within Newton's framework.

In particular, we will adopt a simple model for frictional forces that is specified by two constants for each pair of surfaces that are in contact with each other. These constants are the coefficients of static and kinetic friction. We will then address specific examples to demonstrate how we can use these constants to account for frictional forces within Newton's framework.

6.2 Friction

In the last unit we introduced several new forces. The gravitational force is a fundamental force of nature that exists between any two objects of any size that have mass. In this unit, we will introduce a force, friction, which is more similar to the contact forces we discussed in the last unit.

In general the force between two surfaces that are in contact has components both perpendicular and parallel to the surfaces. The perpendicular component is the normal force we discussed last time. The parallel component is called the **friction force**.

The direction of the friction force is *always* such that it opposes any relative motion of the two surfaces. We distinguish between two kinds of friction forces. **Kinetic friction** refers to cases in which one surface moves relative to the other one, such as when a box slides across the floor, while **static friction** refers to cases in which the surfaces do not move relative to each other, such as when a person is pushing on a stationary heavy box.

The microscopic origins of these friction forces are complicated; they arise from the interactions of atoms on the surface of materials. We will not try to understand these interactions in this course. We will only be concerned with characterizing the friction forces on macroscopic objects using a simple model.

6.3 Kinetic Friction

When one object slides across another the frictional force between them is found to depend linearly on the perpendicular force between them. In other words, the frictional force is proportional to the normal force.

$$f_k = \mu_k N$$

The constant of proportionality, μ_k, is called the **coefficient of kinetic friction**. This constant depends only on the properties of the two surfaces and not on the size or weight of the objects.

We'll now do an example to see how this works. Suppose a box is given an initial shove after which it slides on a horizontal floor. If the coefficient of kinetic friction between the floor and the box is μ_k, what is the acceleration of the box as it slows to a stop?

We will follow the problem solving procedure we developed last time. We first draw the free-body diagram for the box as shown in Figure 6.1. The forces acting on the box once it is in motion are the weight of the box, the normal force exerted by the ground on the box, and the kinetic friction force. The next step is to write down Newton's second law for both the x and y directions.

$$\sum F_x = ma_x \qquad \rightarrow \qquad -\mu_k N = ma_x$$

$$\sum F_y = ma_y \qquad \rightarrow \qquad N - mg = 0$$

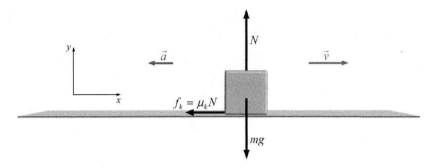

FIGURE 6.1 The free-body diagram for a box of mass m sliding across a floor characterized by a coefficient of sliding friction μ_k between the box and the floor.

The acceleration in the y direction is zero; therefore, the magnitude of the normal force is just equal to the weight of the box. Turning now to the x direction, we see that we can determine the acceleration since we now know the magnitude of the kinetic friction force. Namely, the acceleration in the x direction is just equal to the coefficient of kinetic friction times the acceleration due to gravity.

$$a_x = -\mu_k g$$

The minus sign indicates that the direction of the acceleration is to the left.

Note that the acceleration does not depend upon the mass of the box! How does this come about? The reason is that the net force here is just the kinetic friction force which, by its definition, is proportional to the normal force. But the normal force here is equal to the weight of the box; therefore, we see that the kinetic friction force is proportional to the mass of the box. Consequently, when we write down Newton's second law, we see the mass cancels. Therefore, we see that the box slows down with a constant acceleration whose magnitude depends only on the coefficient of kinetic friction.

6.4 Static Friction

We will now turn to another everyday example that illustrates how we account for static friction. Figure 6.2 shows a box is at rest on a horizontal floor being pulled gently by a rope attached to it. The box does not move. Pull a bit harder, and the box still does not move. If we keep pulling harder and harder, eventually the box does move. How do we describe what happened in terms of Newton's laws? Clearly the acceleration of the box was zero until it began to move. We know then from Newton's second law, that until the box began to move, the total force on the box was zero! Since the rope was exerting a force on the box, directed to the right, there must have been another force of the same magnitude that was directed to the left. This force is the static friction force $\vec{f}$; it opposes the horizontal motion that the tension force would have produced.

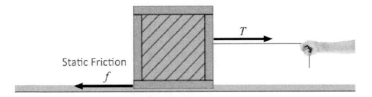

FIGURE 6.2 A rope exerts a force on a box and the floor exerts a frictional force on the box. The box is stationary, implying that the magnitude of the static frictional force is equal to the magnitude of the tension force.

The magnitude of the tension force reflects how hard we pull. The harder we pull, the bigger the tension gets, and therefore, the bigger the frictional force becomes. This dependence of the frictional force on the tension is demonstrated in the plot shown in Figure 6.3.

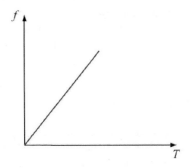

FIGURE 6.3 A plot of the magnitude of the frictional force as a function of the tension of the rope used in the example in Figure 6.2.

There is a limit to the force that friction can provide, though, and eventually we are able to move the box just by pulling harder. This maximum value of the force that static friction can provide is once again just proportional to the normal force. In this case the constant of proportionality is called the **coefficient of static friction**, μ_s, and depends only on the properties of the two surfaces. This behavior of the static friction force that we have described is captured in an inequality; namely, that the static friction force is *less than or equal to* the coefficient of static friction times the normal force.

$$f_s \leq \mu_s N$$

Note the important difference between kinetic and static friction: the kinetic force is always equal to $\mu_s N$, while the static force is NOT always equal to $\mu_s N$. In fact, the static force is only equal to $\mu_s N$ just before the surfaces began to move. In all other cases, the static force is less than this maximum force; just as it was for the normal force and for the tension force, the magnitude of the static frictional force must be determined from Newton's second law: *The static frictional force is simply what is has to be to do what it does!*

We can summarize what we have learned about friction so far by completing the plot we have created for static friction (Figure 6.4). As we increase $\vec{T}$ beyond the point at which the box begins to move, we see that frictional force stays constant as $\vec{T}$ increases. The discontinuity at the point at which the box begins to move indicates that $\mu_s > \mu_k$.

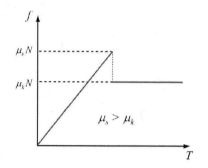

FIGURE 6.4 A plot of the magnitude of the frictional force as a function of the tension of the rope used in the example in Figure 6.2. The force is discontinuous at the point where the box begins to move.

6.5   Example: Box Sliding Down a Ramp

We will now do some examples that illustrate the use of these frictional forces within the framework of Newton's laws. We will start with the calculation of the acceleration of a box as it slides down a rough ramp.

Figure 6.5 shows a free-body diagram of all of the forces acting on the box. In the absence of friction between the ramp and the box, there are two forces acting on the box: its weight (mg), which points downward, and the normal force (N) that the ramp exerts on the box. The direction of this normal force is perpendicular to the ramp. Right away we can see that there is a net force directed down the plane which gives rise to the acceleration down the plane. If there is friction between the box and the ramp, then there will be an

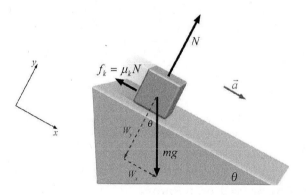

FIGURE 6.5 A free-body diagram for a box of mass m sliding down a ramp. The friction between the ramp and the box is specified by the coefficient of kinetic friction, μ_k.

additional force that opposes this motion. Therefore, this frictional force is directed up the plane and has a magnitude equal to $\mu_k N$, when the box is sliding.

We now need to choose a coordinate system. As we mentioned last time, choosing one axis to be parallel to the acceleration often simplifies the calculation. Therefore, we will choose our x-axis to point down the ramp and the y-axis to be perpendicular to the ramp.

Finally, we write down Newton's second law for both the x and y directions. To write down the y equation, we need to find the y component of the weight.

$$\sum F_y = ma_y \qquad \rightarrow \qquad N - W_y = 0$$

Since the normal force is perpendicular to the ramp and the weight is perpendicular to the horizontal, we see that the angle between these two forces must also be equal to θ, the angle the ramp makes with the horizontal. Therefore, we can use trigonometry to determine that the y component of the weight is just equal to the weight times the cosine of the angle the ramp makes with the horizontal ($W_y = mg \cos \theta$). Substituting this into Newton's second law for the y direction, we obtain the magnitude of the normal force. This result holds for both the friction and frictionless cases.

$$N = mg \cos \theta$$

We can now obtain a value for the magnitude of the frictional force from our result for the magnitude of the normal force. The x component of the weight is equal to the product of the weight and the sine of the angle the ramp makes with the horizontal.

We can now write down Newton's second law in the x direction in terms of this frictional force and the component of the weight down the ramp.

$$\sum F_x = ma_x \qquad \rightarrow \qquad mg \sin \theta - \mu_k N = ma_x$$

We can substitute our result for the normal force into this equation and then solve for the acceleration. We obtain the result that the acceleration is equal to the g times a factor that is determined by the coefficient of kinetic friction and the angle the ramp makes with the horizontal.

$$a = g(\sin \theta - \mu_k \cos \theta)$$

We can remove the frictional force by setting μ_k equal to zero, to determine that the acceleration in the absence of friction is just equal to the product of g and the sine of the angle the ramp makes with the horizontal.

6.6   Example: Box at Rest on a Ramp

We have just determined the acceleration of a box sliding down a ramp. In particular, we see that the acceleration decreases as the angle θ decreases. If we make the angle small enough, the frictional force on the box will prevent it from accelerating at all! At such an angle, what is the magnitude of the frictional force?

To answer this question, we start with the free-body diagram shown in Figure 6.6.

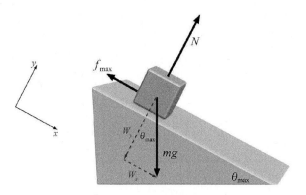

FIGURE 6.6 A free-body diagram for a box of mass m on a ramp whose angle with the horizontal, θ_{max}, is the largest it can be to prevent the box from sliding down the ramp.

We see that this diagram is essentially identical to that in Figure 6.5, as long as we just call the frictional force $\vec{f}$. Since the box is not moving, we have a case of static friction. The magnitude of the static friction force must be determined from Newton's second law. In particular, the magnitude of the static friction force must be equal to the component of the weight down the plane.

$$\sum F_x = mg \sin \theta - f = 0$$

As we increase θ, both the friction force and the x component of the weight increase. At some angle θ_{max}, though, the static friction force can increase no more and the block will begin to slide. We can determine θ_{max} by setting the static friction force equal to its maximum value ($\mu_s N$).

$$f_{max} = \mu_s N = \mu_s mg \cos \theta_{max}$$

Substituting this value for the frictional force back into our general equation,

$$mg \sin \theta_{max} - \mu_s mg \cos \theta_{max} = 0$$

we obtain our result for the maximum angle the ramp can make with the horizontal to prevent the box from sliding down the ramp.

$$\tan \theta_{max} = \mu_s$$

6.7   Example: Car Rounding a Corner

We will close this unit by doing one more example involving friction, namely, that of a car of mass m rounding a circular turn of radius R. If the coefficient of static friction between the tires and the road is μ_s, how fast can the car go around the turn without skidding off the road?

Perhaps your first question here is why in the world are we giving you the static coefficient of friction when the car is clearly moving? The answer to this question is that the tires are rolling: the surfaces of the tires are not sliding relative to the surface of the road, since if they were, the car would already be skidding! During normal driving it is the static friction between the tires and the road that makes a car speed up, slow down, and turn corners!

We start, as always, by drawing a free-body diagram for the car, as shown in Figure 6.7. From this diagram, we see it is just the frictional force $\vec{f}$ that is responsible for the centripetal acceleration of the car. Next, we choose the x-axis to point toward the center of the circle to align it with the direction of the acceleration. We choose the y-axis to be vertically up.

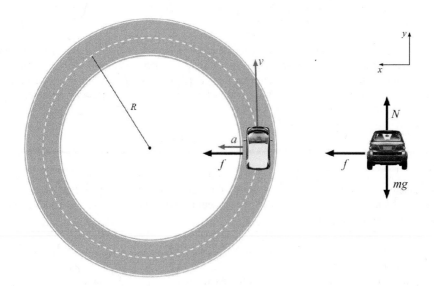

FIGURE 6.7 The free-body diagram for a car of mass m moving at constant speed v in a circle of radius R.

Writing Newton's second law for the y direction, we see that the magnitude of the normal force is just equal to the weight of the car.

$$\sum F_y = N - mg = ma_y = 0$$

Writing Newton's second law for the x direction, we see that the frictional force must be equal to the mass of the car times the centripetal acceleration.

$$\sum F_x = f = ma_x$$

We know the centripetal acceleration is equal to the square of the speed divided by the radius of the turn.

$$a_x = \frac{v^2}{R}$$

Therefore, as the car's speed increases, the frictional force must also increase. There is a limit though as to how much this frictional force can increase. This limit is determined by the coefficient of static friction, μ_s. This maximum frictional force ($f_{max} = \mu_s N = \mu_s mg$) then produces the maximum centripetal acceleration ($a_{max} = v_{max}^2 / R$) which determines the maximum velocity. Therefore, we conclude the maximum speed that the car can make the turn without skidding is proportional to the square root of the product of the turning radius and the coefficient of static friction.

$$v_{max} = \sqrt{a_{max}R} = \sqrt{\frac{f_{max}}{m}R} = \sqrt{\mu_s gR}$$

This result makes sense: the car's maximum speed should increase if either the friction increases or the turning radius increases.

MAIN POINTS

Friction Forces

The frictional force refers to the parallel component of the contact force between two surfaces.

Kinetic Friction

Kinetic friction exists between surfaces in relative motion.

The direction of kinetic friction always opposes the relative motion.

The magnitude of kinetic friction is proportional to the normal force.

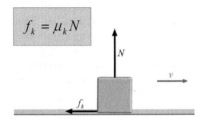

$$f_k = \mu_k N$$

Static Friction

Static friction exists between surfaces not in relative motion.

The direction of static friction always opposes the relative motion that would exist in the absence of friction.

The magnitude of static friction must be determined from Newton's second law. Its maximum value is proportional to the normal force.

The static frictional force is what it has to be to do what it does!

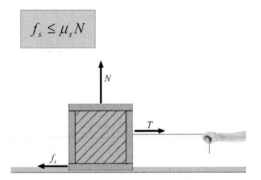

$$f_s \leq \mu_s N$$

PROBLEMS

1. Accelerating Blocks: Two wooden crates rest on top of one another. The smaller top crate has a mass of $m_1 = 25$ kg and the larger bottom crate has a mass of $m_2 = 89$ kg. There is no friction between the bottom crate and the floor, but the coefficient of static friction between the two crates is $\mu_s = 0.86$ and the coefficient of kinetic friction between the two crates is $\mu_k = 0.69$. A massless rope is attached to the lower crate to pull it horizontally to the right (which should be considered the positive direction for this problem). (a) The rope is pulled with a tension $T = 433$ N (which is small enough that the top crate will not slide). What is the acceleration of the small crate? (b) In the previous situation, what is the frictional force the lower crate exerts on the upper crate? (c) What is the maximum tension that the lower crate can be pulled before the upper crate begins to slide? (d) The tension is increased in the rope to 1,356 N causing the boxes to accelerate more and the top box to begin sliding. What is the acceleration of the upper crate? (e) As the upper crate slides, what is the acceleration of the lower crate?

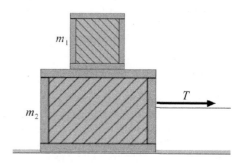

FIGURE 6.8 Problem 1

2. Accelerating Truck: A box rests on top of a flat bed truck. The box has a mass of $m = 18$ kg. The coefficient of static friction between the box and truck is $\mu_s = 0.86$ and the coefficient of kinetic friction between the box and truck is $\mu_k = 0.68$. (a) The truck accelerates from rest to $v_f = 18$ m/s in $t = 15$ s (which is slow enough that the box will not slide). What is the acceleration of the box? (b) In the previous situation, what is the frictional force the truck exerts on the box? (c) What is the maximum acceleration the truck can have before the box begins to slide? (d) Now the acceleration of the truck remains at the value as calculated in (c), and the box begins to slide. What is the acceleration of the box? (e) With the box still on the truck, the truck attains its maximum velocity. As the truck comes to a stop at the next stop light, what is the magnitude of the maximum deceleration the truck can have without the box sliding?

3. Mass on Incline: A block with mass $m_1 = 8.6$ kg is on an incline with an angle $\theta = 30°$ with respect to the horizontal. For the first question there is no friction, but for the rest of this problem the coefficients of friction are $\mu_k = 0.27$ and $\mu_s = 0.297$. (a) When there is no friction, what is the magnitude of the acceleration of the block? (b) With friction, what is the magnitude of the acceleration of the block after it begins to slide down the plane? (c) To keep the mass from accelerating, a spring is attached. What is the minimum

FIGURE 6.9 Problem 3 part (c)

spring constant of the spring to keep the block from sliding if it extends $x = 0.15$ m from its unstretched length. (d) A new block with mass $m_2 = 14.1$ kg is attached to the first block. The new block is made of a different material and has a greater coefficient of static friction. What minimum value for the coefficient of static friction is needed between the new block and the plane to keep the system from accelerating?

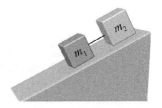

FIGURE 6.10 Problem 3 part (d)

4. Carnival Ride: In a classic carnival ride, patrons stand against the wall in a cylindrically shaped room. Once the room gets spinning fast enough, the floor drops from the bottom of the room! Friction between the walls of the room and the people on the ride make them "stick" to the wall so they do not slide down. In one ride, the radius of the cylindrical room is $R = 6.9$ m and the room spins with a frequency of 20.4 revolutions per minute. (a) What is the speed of a person "stuck" to the wall? (b) What is the normal force of the wall on a rider of $m = 48$ kg? (c) What is the minimum coefficient of friction needed between the wall and the person? (d) If a new person with mass 96 kg rides the ride, what minimum coefficient of friction between the wall and the person would be needed? (e) Which of the following changes would decrease the coefficient of friction needed for this ride: *increasing the rider's mass, increasing the radius of the ride, increasing the speed of the ride*, or *increasing the acceleration due to gravity (e.g., taking the ride to another planet)*? (f) To be safe, the engineers making the ride want to be sure the normal force does not exceed 2.2 times each person's weight, and therefore they adjust the frequency of revolution accordingly. What is the minimum coefficient of friction now needed?

5. Block (INTERACTIVE EXAMPLE): A wooden block of mass $m = 9$ kg starts from rest on an inclined plane sloped at an angle θ from the horizontal. The block is originally located 5 m from the bottom of the plane. If the block, undergoing constant acceleration down the ramp, slides to the bottom in $t = 2$ s, and $\theta = 30°$, what is the magnitude of the kinetic frictional force on the block?

UNIT

7

WORK AND KINETIC ENERGY

7.1 Overview

This unit introduces two important new concepts: *kinetic energy* and *work*. These concepts are defined in terms of the fundamental concepts from dynamics (*force* and *mass*) and kinematics (*displacement* and *velocity*). We will find that integrating Newton's second law through a displacement will result in an equation that links these two concepts of kinetic energy and work. This equation (called the work-energy theorem or sometimes, the center-of-mass equation), allows us to easily answer many questions that would be very difficult using Newton's second law directly.

7.2 Work and Kinetic Energy in One Dimension

We begin our introduction of work and energy by considering the simple one-dimensional situation shown in Figure 7.1.

FIGURE 7.1 A constant force $\vec{F}_{Net}$ is applied to an object over a time interval (from t_1 to t_2) that results in a change in its velocity and its displacement.

A force is applied to an object, causing it to accelerate. We say that the force acting over time causes the change in velocity.

$$F_{Net} = ma \qquad \rightarrow \qquad F_{Net} = m\frac{dv}{dt}$$

We can quantify this statement by integrating the force over time to obtain the relationship between this integral of the force over time and the change in velocity.

$$\int_{t_1}^{t_2} F_{Net}\, dt = m\int_{v_1}^{v_2} dv$$

We could also describe this situation by saying that the force acting through a *distance* caused the change in velocity. How do we quantify this description? Well, consider the motion at time t: in the next instant of time dt, the velocity will change by an amount dv which is equal to the acceleration times dt.

$$dv = a\, dt$$

In this same time dt, the position will change by an amount which is equal to the velocity times dt.

$$dx = v\, dt$$

We apply Newton's second law, to replace the acceleration by the net force divided by the mass and then combine the equations to eliminate dt and obtain the equation:

$$dv = \frac{F_{Net}}{m}\frac{dx}{v}$$

If we now integrate this equation, we obtain the relationship between the integral of the net force over the displacement and the change in the square of the velocity.

$$m\int_{v_1}^{v_2} v\, dv = \int_{x_1}^{x_2} F_{Net}\, dx \qquad \rightarrow \qquad \frac{1}{2}m\Delta(v^2) = \int_{x_1}^{x_2} F_{Net}\, dx$$

We define the integral of the force over the displacement to be the **work** done by the force:

$$W \equiv \int_{x_1}^{x_2} F\, dx$$

The quantity one-half the mass times the velocity squared is the **kinetic energy** of the particle:

$$K \equiv \frac{1}{2}mv^2$$

Thus, we see that in one dimension, the work done on the object by the net force is equal to the change in that object's kinetic energy. Work and kinetic energy both are measured in joules (J), where 1 joule is defined to be 1 N·m. We will do a simple example in the next section that illustrates the use of these concepts.

7.3 Example

Figure 7.2 shows a box of mass 6 kg that is initially at rest. A horizontal force of magnitude 24 N is now applied and the box begins to move. We would like to determine the speed of the box when it is at a distance of 8 m from its initial position.

How do we go about making this calculation? We could first use Newton's second law to determine the acceleration, then use this acceleration in one of our kinematics equations to determine the time it takes to travel 8 m, and finally use another kinematics equation to

FIGURE 7.2 A constant force of 24 N is applied to a box of mass 6 kg that is initially at rest. What is the speed of the box when it reaches a distance of 8 m from its starting point?

determine the speed at this time.

There is, however, an easier way. Namely, we can simply equate the work done by the net force (which is just the applied force in this case) to the change in the kinetic energy of the box.

$$W = \Delta K$$

Now, since the applied force is constant, we can take it outside the integral, and then the work done is just equal to the product of the applied force and the distance. Since the box was initially at rest, the change in kinetic energy of the box is just equal to its final kinetic energy.

$$W = F \cdot (x_f - x_i) = (24\,\text{N})(8\,\text{m})$$

$$\Delta K = \frac{1}{2}mv_f^2 = \frac{1}{2}(6\,\text{kg})v_f^2$$

Consequently, we see that the final speed is equal to 8 m/s.

We've just seen how helpful this connection between the work done and the change in kinetic energy in one dimension can be. In order to generalize this connection to more than one dimension, we will need to introduce the concept of the dot product of two vectors, which we will do in the next section.

7.4   The Dot Product

So far, all operations we have performed on vectors have produced another vector. When we add or subtract vectors, the result is another vector. When we multiply a scalar by a vector, the result is also a vector.

How do we multiply two vectors? There are actually two different products of two vectors. The cross product of two vectors produces another vector, and we will discuss this operation later in the course. In this section we will introduce the *dot product* of two

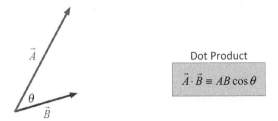

Dot Product

$$\vec{A} \cdot \vec{B} \equiv AB \cos \theta$$

FIGURE 7.3 The dot product of two vectors is defined to be a scalar that measures the projection of one vector along the other.

vectors which produces a scalar.

The **dot product** of any two vectors is defined to be the product of the magnitudes of the vectors and the cosine of the angle between them as shown in Figure 7.3.

Therefore, if the two vectors are parallel, the dot product is equal to the arithmetic product.

If the two vectors are perpendicular, the dot product is equal to zero. If the two vectors are anti-parallel, the dot product is equal to minus the arithmetic product. The dot product of two vectors is a measure of the projection of one vector along the other.

The dot product is used to define the components of vectors. For example, A_x and A_y are the dot products of $\vec{A}$ with the unit vectors in the x and y directions, respectively.

Figure 7.4 shows two vectors $\vec{A}$ and $\vec{B}$, defined in terms of their x and y components. If we now take the dot product of these two vectors, we see that the only terms that survive are the products of the same components

$$\vec{A} \cdot \vec{B} = A_x B_x + A_y B_y$$

In the next section, we will use the dot product to define the work done by a force.

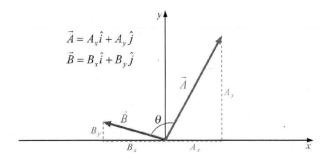

FIGURE 7.4 The representation of vector $\vec{A}$ and $\vec{B}$ in terms of their components (A_x, A_y) and (B_x, B_y) using the dot product and the unit vectors ($\hat{i}, \hat{j}$) along the (x, y) axes.

7.5   The Work-Kinetic Energy Theorem

In an earlier section we determined that in one dimension, the integral of the net force over the displacement of an object is proportional to the change in the square of its velocity We will now make use of the dot product to generalize this result to more than one dimension.

We start by defining the work done by a force on an object as the integral of the dot product of the force and the displacement.

$$W \equiv \int \vec{F} \cdot d\vec{l}$$

We can see from this definition that only the part of the force that is parallel to the displacement contributes to the calculation of the work done by the force. For example if you pull a box across a horizontal frictionless floor, both the weight and the normal force are perpendicular to the displacement so that neither of these forces does any work on the box. Only the tension force has a component in the direction of the displacement, so only the tension force does any work on the box.

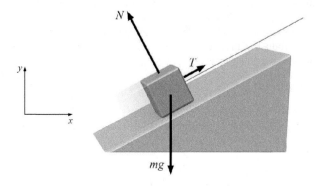

FIGURE 7.5 A free-body diagram for a box of mass m being pulled up a ramp by a taut rope.

Figure 7.5 shows a box being pulled up a frictionless ramp so that we have displacements in two dimensions, x and y. We can now basically repeat the derivation of the one dimensional case for each component. That is, we start from the equations for the change in velocity and displacement during a time dt.

$$dv_x = a_x \, dt \qquad\qquad dx = v_x \, dt$$
$$dv_y = a_y \, dt \qquad\qquad dy = v_y \, dt$$

Once again we apply Newton's second law to replace the acceleration by the net force divided by the mass and combine the equations to eliminate dt and obtain two equations, one for each dimension.

$$mv_x \, dv_x = F_{Net,x} \, dx \qquad\qquad mv_y \, dv_y = F_{Net,y} \, dy$$

We now integrate each equation and add them to obtain our final result:

$$m\left(\int v_x \, dv_x + \int v_y \, dv_y \right) = \int (F_{Net,x} \, dx + F_{Net,y} \, dy)$$

The left-hand side of the equation is equal to the change in kinetic energy of the box. The right-hand side of the equation is the work done by the net force, since the dot product can be expanded as the sum of the products of the components.

$$\Delta K = W_{Net}$$

We now see that defining the work in terms of the dot product of the force and the displacement allows us to generalize our one-dimensional result. *The change in the kinetic energy of an object is equal to the work done on that object by the net force.* We call this statement the **work-kinetic energy theorem**. In the next few sections we will explore the work done by several forces.

7.6 Examples: Work Done by Gravity Near the Surface of the Earth

Figure 7.6(a) shows a mass m being moved along some arbitrary path connecting two fixed points near the surface of Earth. We want to calculate the work done by gravity along this path. We will first approximate the path as a series of infinitesimal straight line segments along the horizontal and vertical directions as shown in Figure 7.6(b).

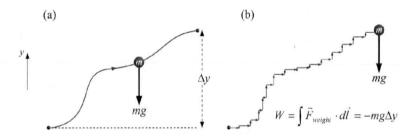

FIGURE 7.6 (a) An object of mass m is moved along an arbitrary path between two fixed points separated by a vertical distance Δy. (b) The work done by the weight force is calculated by approximating the path by a series of infinitesimal horizontal and vertical segments. The only non-zero contributions come from the vertical segments.

The work done along all of the horizontal segments is zero since the force (the weight) is always perpendicular to the direction of the horizontal segment. Therefore, the total work done by gravity along this path is just the sum of the work done by gravity along the vertical segments. Since the force (the weight) points down and the vertical segment (dy) points up, the dot product of these two vectors is just equal to $-mg\,dy$. Thus, the total work done by gravity along the path is equal to minus the product of the weight and the total vertical displacement.

$$W = -mg\,\Delta y$$

Note that this formula for the work done by the weight does *not* depend on the path taken but only on the difference in height of the initial and final points. When the work done by a force during some motion only depends on the endpoints of the motion, but not on the details of the path, we say that the force is **conservative**. A consideration of conservative forces will lead us to the important concept of potential energy in the next unit.

We will now use the result for the work done by the weight in an example that illustrates the power of the work-kinetic energy theorem. Figure 7.7 shows two balls released at the same time from the same height h. One ball simply falls with constant acceleration g. The other skids down a frictionless curved surface. The free fall problem is one we've solved many times; we know the velocity and position of the ball at any time. We use this information to determine how long the ball is in the air and with what velocity it hits the ground. The skidding ball looks to be a much more difficult problem.

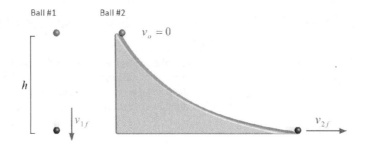

FIGURE 7.7 Two balls are dropped at the same time from the same height. Ball 1 falls freely while ball 2 skids down a frictionless curved surface.

The net force here is certainly not constant! Therefore, we cannot use our kinematic equations for constant acceleration. Indeed, finding the position and velocity of the skidding ball at any time is difficult. However, we can use the work-kinetic energy theorem to find easily the final velocity.

Only two forces act on the skidding ball: the normal force provided by the surface and the weight of the ball. The ball always moves parallel to the plane. The normal force is always perpendicular to the plane. Therefore, work done by the normal force, given by the dot product of the normal force and the displacement, is zero! Consequently, the work done by the net force here is just equal to the work done by the weight:

$$W = -mg\,\Delta y$$

Applying the work-kinetic energy theorem, we see that the final speed of the skidding ball is exactly the same as the final speed for the dropped ball!

$$-mg\Delta y = \frac{1}{2}mv_{2f}^2$$

$$-mg(-h) = \frac{1}{2}mv_{2f}^2$$

$$v_{2f} = \sqrt{2gh}$$

We will close this unit with a discussion of two more examples of the work done by conservative forces.

7.7   Work Done by a Variable Force: The Spring Force

We have calculated the work done by the weight, a constant force, when the orientation of the path relative to the force was changing. We will now calculate the work done by a spring in one dimension. In this case, the orientation of the path relative to the force is simple, but the magnitude of the force changes as we move.

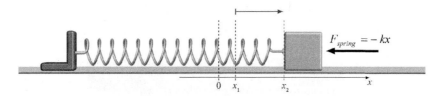

FIGURE 7.8 A spring with a spring constant k exerts a restoring force on an object of mass m. The coordinate system is defined such that $x = 0$ corresponds to the relaxed length of the spring. To calculate the work done by the spring as the object moves between two positions, x_1 and x_2, we must integrate the varying restoring force from x_1 to x_2.

We define the origin, or the coordinate system, to correspond to the relaxed length of the spring as shown in Figure 7.8. The force exerted by the spring on the object attached to its end as a function of position is proportional to the extension or compression of the spring. As we move the object between two positions x_1 and x_2, the force on the object clearly changes. Breaking the movement into tiny steps we see that the work done by the spring along each step will depend on the position. To find the total work done, we need to integrate this expression between x_1 and x_2.

$$W_{1 \to 2} = -k \int_{x_1}^{x_2} x \, dx = -\frac{1}{2} k (x_2^2 - x_1^2)$$

Figure 7.9 shows a plot of the force as a function of displacement as the spring is moved from x_1 to x_2. The area under the curve represents the work done by the force.

Note that the formula for the work done by a spring that we just derived depends only on the endpoints of the motion, x_1 and x_2. In fact, if we first stretch the spring from its

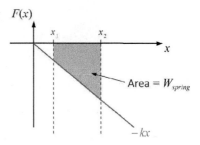

FIGURE 7.9 A plot of the force exerted by the spring in Figure 7.8 as a function of the extension of the spring. The work done by the spring as the object is moved from x_1 to x_2 is given by the shaded area under the curve.

equilibrium position, we see that the spring force points in the opposite direction to the displacement and therefore does negative work. If we then compress the spring back to its equilibrium position, the spring force has the same direction, but the displacement is now in the opposite direction so that the spring force does positive work. In fact the magnitudes of these negative and positive works are equal, so that we see that when the spring is returned to its original position, the net work done is zero, the signature of a conservative force.

7.8   Work Done by Gravity Far from the Earth

As a final example, we will consider the work done by gravity on an object of mass m that moves in three dimensions along some arbitrary path between two fixed points $\vec{r}_1$ and $\vec{r}_2$ that are far from the surface of Earth as shown in Figure 7.10. In this case both the direction of the path relative to the direction of the force, as well as the magnitude of the force, can change along the path.

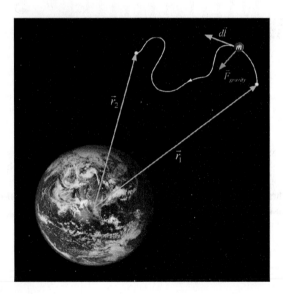

FIGURE 7.10 To calculate the work done by gravity as a mass m is moved along the curved path shown, we need to evaluate an integral in which the magnitude of the force $\vec{F}_{gravity}$ is changing as is the angle between the force and the path element.

The magnitude of the gravitational force that the Earth exerts on this object is given by the universal gravitational force law; the direction of this force is always radially inward, toward the center of the Earth. Once again, we can break the path between $\vec{r}_1$ and $\vec{r}_2$ into tiny steps as shown in Figure 7.11. One such step is shown magnified in the figure. This step is in turn broken into two even smaller steps, one that is parallel to the radial direction and one which is perpendicular to the radial direction. Since the force is radial, the work done when moving along the perpendicular step is zero! The work done when moving along the radial step is just equal to the force at that point times the radial displacement.

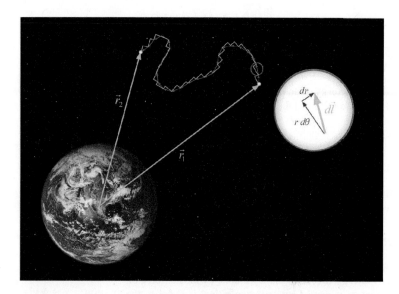

FIGURE 7.11 To evaluate the work done by gravity, we approximate the curved path as a series of infinitesimal elements each of which are in turn broken down into two elements, one radial and one tangential. The work done along each tangential element is zero since the force is radial. Consequently, the work done by gravity only depends on r_1 and r_2, the radial distances of the endpoints of the path.

$$dW = \vec{F}_{gravity} \cdot d\vec{l} = -G\frac{M_E m}{r^2}\hat{r} \cdot d\vec{l} = -G\frac{M_E m}{r^2}dr$$

The total work is therefore just the sum of the work done by all of the radial steps.

To make this sum, we do the integral and obtain our result: the work done by the gravitational force as an object is moved from $\vec{r}_1$ to $\vec{r}_2$ is proportional to $1/r_2 - 1/r_1$.

$$W_{1\rightarrow2} = -\int_{r_1}^{r_2}\frac{GM_E m}{r^2}dr = GM_E m\left(\frac{1}{r_2} - \frac{1}{r_1}\right)$$

Once again we see that the work depends only on the endpoints and not on the specific path, demonstrating that the gravitational force is indeed a conservative force.

MAIN POINTS

Work and Kinetic Energy Definitions

The **kinetic energy** of an object is defined to be ½ the product of the mass and the square of its velocity.

$$K \equiv \frac{1}{2}mv^2$$

The **work** done by a force as an object is moved between two points along some path is defined to be the dot product of the force with the displacement along the path between those two points.

$$W_{1\to2} \equiv \int_{\vec{r}_1}^{\vec{r}_2} \vec{F} \cdot d\vec{l}$$

The Work-Kinetic Energy Theorem

Integrating Newton's second law, we obtain the **work-kinetic energy theorem**: the work done by the net force on an object as it moves between two points is equal to the change in its kinetic energy.

$$W_{Net} = \Delta K$$

Conservative Forces

Conservative forces are defined to be those forces in which the work done by them does NOT depend on the path, only on the endpoints.

Work Done by Two Conservative Forces

Gravitational Force	$W = -mg\,\Delta y$	(near Earth)
	$W_{1\to2} = GM_E m\left(\dfrac{1}{r_2} - \dfrac{1}{r_1}\right)$	(general expression)

Spring Force	$W_{1\to2} = -\dfrac{1}{2}k(x_2^2 - x_1^2)$	

PROBLEMS

1. Work on Two Blocks: A mass $m_1 = 4.2$ kg rests on a frictionless table and is connected by a massless string over a massless pulley to another mass $m_2 = 4.8$ kg, which hangs freely from the string. When released, the hanging mass falls a distance of $d = 0.72$ m. (a) How much work is done by gravity on the two block system? (b) How much work is done by the normal force on m_1? (c) What is the final speed of the two blocks? (d) How much work is done by tension on m_1? (e) What is the tension in the string? (f) Is the work done by tension on only m_2 *positive*, *zero*, or *negative*? (g) What is the net work done on m_2?

FIGURE 7.12 Problem 1

2. Block Sliding: A block with mass $m = 13$ kg rests on a frictionless table and is accelerated by a spring with spring constant $k = 4,509$ N/m after being compressed a distance $x_1 = 0.454$ m from the spring's unstretched length. The floor is frictionless except for a rough patch a distance $d = 2.6$ m long. For this rough path, the coefficient of friction is $\mu_k = 0.4$.

FIGURE 7.13 Problem 2

(a) How much work is done by the spring as it accelerates the block? (b) What is the speed of the block right after it leaves the spring? (c) How much work is done by friction as the block crosses the rough spot? (d) What is the speed of the block after it passes the rough spot? (e) Instead, the spring is only compressed a distance $x_2 = 0.114$ m before being released. How far into the rough path does the block slide before coming to rest? (f) What distance does the spring need to be compressed so that the block will just barely make it past the rough patch when released? (g) If the spring was compressed three times farther and then the block is released, is the work done on the block by the spring as it accelerates the block *the same*, *three times greater*, *three times less*, *nine times greater*, or *nine times less*?

CONSERVATIVE FORCES
AND POTENTIAL ENERGY

8.1 Overview

This unit introduces an important new concept: potential energy. In particular, for any conservative force, we can define the change in potential energy of an object as minus the work done by this force. In this course, we deal with two conservative forces, gravity and springs. After defining the potential energy associated with each of these forces, we can rewrite the work-kinetic energy theorem so that it expresses a conservation law: the conservation of mechanical energy that applies whenever the only forces that do work in the situation are conservative forces.

8.2 Conservative Forces

In the last unit we introduced the concept of work as a force acting over some distance, and we showed that work done on an object will change its kinetic energy. We evaluated the work done by two of the forces we have discussed so far, gravity and springs, and we found that for these forces, the work done on an object by these forces depends only on the starting and ending points of the motion, and not on the path taken between these points.

In general, forces that have the property that the work done by them is independent of the path used to integrate between the two points are called **conservative forces**. If the work done by a force between two points is independent of the path, then it must be true that the work done by such a force on a closed path, that is to say a path that ends up where it started, is zero. To prove this claim, consider any two points x_1 and x_2 on the closed path as shown in Figure 8.1.

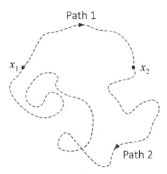

The work done from x_1 to x_2 along the top segment must be equal to the work done from x_1 to x_2 on the bottom segment, since the force is conservative. However, the work done from x_1 to x_2 on the bottom segment must be equal to minus the work done from x_2 to x_1 along the bottom segment, since we have just interchanged the endpoints. We can see that the work done around the closed loop

FIGURE 8.1 The work done by a conservative force around any closed path is zero. For example, the work done from x_1 to x_2 along Path 1 is equal to minus the work done from x_2 to x_1 along Path 2.

is equal to the sum of the work done from x_1 to x_2 along the top segment and the work done from x_2 to x_1 along the bottom segment. This sum is zero!

8.3 Potential Energy

In general, the work done by a force on an object between two points does depend on the path taken by the object between the two points. For the special case of conservative forces, we have seen that the work does not depend on the path. Therefore, we can define, for conservative forces, an associated **potential energy** that, for a given object, depends only on its location. In particular, when a conservative force acts on an object as it moves between two points, we define the *change* in potential energy associated with that force as negative the work done by that force between those two points.

$$\Delta U = U_B - U_A \equiv -W_{A \to B}$$

At this point, this definition of potential energy must seem quite arbitrary to you. If we look at this definition in the context of the work-kinetic energy theorem, however, it will begin to make sense.

Recall the example from the last unit (shown in Figure 8.2) in which we applied the work-kinetic energy theorem to determine that the speed that a ball, released from rest, attains while sliding down a frictionless surface only depends on the change in height of the ball

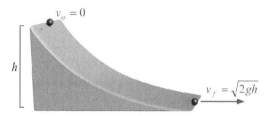

FIGURE 8.2 The speed of the ball at the bottom of the frictionless surface depends only on h, the change in height of the ball.

and not on the details of the surface. In particular, we found that the change in kinetic energy of the ball was equal to the work done by the gravitational force. By our definition, the change in gravitational potential energy is equal to minus the work done by the gravitational force.

$$\Delta U_{gravity} = -W_{gravity} = -(-mg\,\Delta h) = mg\,\Delta h$$

Consequently, we see that the change in potential energy is just equal to the product of the weight of the ball and the change in height between the two points. Therefore, we see that the change in kinetic energy is equal to the negative change in potential energy.

$$\Delta K = W_{gravity} = -\Delta U_{gravity}$$

Whatever kinetic energy the ball gains is equal exactly to the potential energy that the ball loses. If we define the **total mechanical energy** of an object as the sum of its kinetic and potential energy, then, in this case, we see that the total mechanical energy of the ball was conserved. That is, at any point during the motion of the ball, the sum of its kinetic energy plus its gravitational potential energy is a constant.

$$\Delta E_{Mechanical} \equiv \Delta(K + U_{gravity}) = \Delta K + \Delta U_{gravity} = 0$$

8.4   Conservation of Mechanical Energy

In the example shown in Figure 8.2, we demonstrated that the total mechanical energy of the ball, the sum of its kinetic energy and potential energy, was constant throughout its motion. We will now examine the work-kinetic energy theorem to determine exactly when mechanical energy is conserved.

The work-kinetic energy theorem was derived from Newton's second law and states that the change in an object's kinetic energy is equal to the work done by the net force on that object. We can expand the work done by the net force as the sum of the work done by conservative forces and the work done by non-conservative forces.

$$\Delta K = W_{Net} = W_C + W_{NC}$$

In this course, the only *conservative* forces we encounter are the *gravitational force* and the *spring force*. All other forces, for example, friction, tension, etc., are non-conservative forces. If we now move the work done by conservative forces term to the left-hand side of the equation and apply the definition of potential energy, we can see that the sum of the change in kinetic energy and the change in potential energy is equal to the work done by the non-conservative forces.

$$\Delta K - W_C = \Delta K + \Delta U = \Delta E_{Mechanical} = W_{NC}$$

The sum of the change in kinetic energy and the change in potential energy is defined to be the change in the mechanical energy of the object. Therefore, we see that whenever the work done by non-conservative forces is zero, the change in mechanical energy is zero. That is, the *mechanical energy* is *conserved* whenever *the work done by all the non-conservative forces is zero*.

We will encounter many situations in this course in which the work done by the non-conservative forces is zero. In these cases, we can apply the conservation of mechanical energy to answer easily many questions that might be difficult to answer using Newton's laws directly. Indeed, the conservation of mechanical energy gives us the relationship between the location of the object and its speed.

We willl now do a couple of examples that illustrate the power of this conservation of mechanical energy law.

8.5   Gravitational Potential Energy

We've defined the change in potential energy as minus the work done by a conservative force between two points. We can convert this change in potential energy to a potential energy function defined at any single point by simply choosing some specific point as the zero of potential energy. For example, we can define the gravitational potential energy of a mass m near the surface of the Earth as simply mgh, where h is the height of the mass

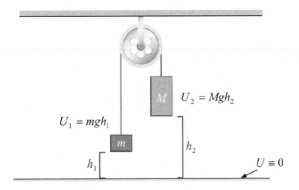

FIGURE 8.3 The gravitational potential energy U can be defined at any point by choosing a convenient height to have $U = 0$.

above a convenient, but arbitrary, point which we can choose to be the zero of potential energy as shown in Figure 8.3.

The table below shows the form of the potential energy function for all conservative forces we will deal with in this course. The arbitrary constant U_o appears in the general form for the potential energy function since it will always cancel when we calculate the physically significant *change* in potential energy.

TABLE 8.1 The force, work done by, and potential energy expressions for the conservative forces we will deal with in this course.

	Force F	Work $W_{1 \rightarrow 2}$	Change in P.E. $\Delta U = U_2 - U_1$	P.E. Function U
Gravity (Near Earth)	$m\vec{g}$	$-mg(h_2 - h_1)$	$mg(h_2 - h_1)$	$mgh + U_o$
Gravity (General Expression)	$-G\dfrac{m_1 m_2}{r^2}\hat{r}$	$Gm_1 m_2 \left(\dfrac{1}{r_2} - \dfrac{1}{r_1} \right)$	$-Gm_1 m_2 \left(\dfrac{1}{r_2} - \dfrac{1}{r_1} \right)$	$-G\dfrac{m_1 m_2}{r} + U_o$
Spring	$-k\vec{x}$	$-\dfrac{1}{2}k(x_2^2 - x_1^2)$	$\dfrac{1}{2}k(x_2^2 - x_1^2)$	$\dfrac{1}{2}kx^2 + U_o$

We will now do an example using the potential energy associated with the universal gravitational force. Suppose we release a ball from a spot far away from Earth and want to know how fast will it be moving when it finally gets here. If the ball is released from rest its change in kinetic energy is proportional to the square of its final speed.

$$\Delta K = \frac{1}{2}m_{ball}v_f^2$$

The change in its potential energy can be found using the expression shown in the table. Since the initial distance is far from the Earth, we can approximate one over the initial distance as zero. What is the final distance? To determine this distance we need to recall that when we discussed the application of Newton's universal gravitational force law between an object and the Earth, we said that we could consider all of the mass of the Earth to be located at its center. Therefore, the final distance here must be equal to the radius of the Earth!

$$\Delta U = -G\frac{M_E m_{ball}}{R_E}$$

Applying the conservation of mechanical energy ($\Delta K + \Delta U = 0$), we obtain the result that the final speed is proportional to the square root of the ratio of the mass of the Earth to its radius.

$$v_f = \sqrt{\frac{2GM_E}{R_E}}$$

It is not too hard to see that we would have arrived at exactly the same answer if we had started by asking another very interesting question: what is the initial speed we need to launch something with from the surface of Earth so that it never returns? In other words, after being launched from Earth, the object slows down and eventually stops when it is infinitely far away. This speed is called the Earth's escape velocity, and when you plug in the numbers you find that it is about 11,200 m/s!

8.6   Vertical Springs

We have already derived an expression for the change in potential energy of a spring.

$$\Delta U_{spring} = \frac{1}{2}k(x^2 - x_o^2)$$

If we choose x_o as the equilibrium length of the spring, we obtain a simple parabola centered on the origin as shown in Figure 8.4.

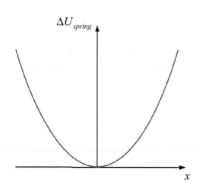

FIGURE 8.4 Defining the potential energy of a spring to be equal to zero at its equilibrium position results in the potential energy function being a parabola.

Suppose we now hang a spring vertically. As long as we assume the spring is massless, it will have the same equilibrium length as before, and the equation for its change in potential energy will have exactly the same form as long as we choose y_o to be its equilibrium length.

Suppose we hang a box of mass m on the end of the spring. This will move the equilibrium position of the system downward to the point where the upward force of the spring balances the weight of the box as shown in Figure 8.5. The amazing thing is that the total change in potential energy of the system due to the spring *and* to gravity combined *still* has the same simple parabolic form as before as long as we make our measurements relative to the new equilibrium position y_e. We will now prove this claim.

Call the displacement from the new equilibrium position y'. We can now write down the expression for the change in potential energy of the spring as we move a distance y' from equilibrium position.

$$\Delta U_{spring} = \frac{1}{2}k[(y' - y_e)^2 - y_e^2]$$

We can simplify this expression to obtain two terms.

$$\Delta U_{spring} = \frac{1}{2}ky'^2 + ky_e y'$$

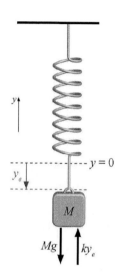

The first term is what we want. We can rewrite the second term by replacing the ky_e factor (the magnitude of the force exerted by the spring) by the weight of the box. Once we make this replacement, we see that the second term is actually equal to minus the change in potential energy due to gravity.

$$ky_e y' = mgy' = -\Delta U_{gravity}$$

Therefore, if we move this second term to the left-hand side of the equation, we obtain the expression that we want:

$$\Delta U_{spring} + \Delta U_{gravity} = \frac{1}{2}ky'^2$$

FIGURE 8.5 A mass M is hung from a spring stretching it a distance y_e from its unstretched length.

Namely, the sum of the change in potential energy due to the spring and the change in the potential energy due to gravity, that is, the change in the *total potential energy* of the system, is equal to the usual expression for the change in the potential energy of a spring if we choose the zero of potential energy to be the equilibrium position when the box is attached!

$$\Delta U_{Total} = \frac{1}{2}ky'^2$$

The beautiful bottom line here is that the change in potential energy of masses hanging from vertical springs is the same simple formula as masses attached to horizontal springs, just as long as we measure the length of the spring *relative* to its *equilibrium length* in both cases.

8.7   Non-Conservative Forces

We will close this unit with a brief discussion of non-conservative forces in the context of the work-kinetic energy theorem and potential energy.

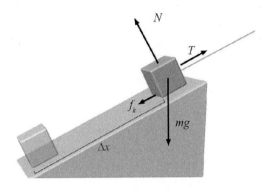

FIGURE 8.6 The forces acting on a box being pulled through a displacement Δx up a ramp are the tension T, the weight mg, the normal force N, and the kinetic friction force f_k.

Figure 8.6 shows a box being pulled up a ramp through a displacement Δx. The forces acting are the weight, the tension, the normal force and the kinetic friction force. If we write down the work-kinetic energy theorem applied to the box as a single rigid object and expand the work done by the net force as the sum of the works done by the individual forces, we obtain:

$$\Delta K = W_{gravity} + W_{tension} + W_{normal} + W_{friction}$$

The only conservative force acting is the weight. We can bring its term to the left-hand side of the equation and call it the change in potential energy.

$$\Delta K + \Delta U_{gravity} = W_{tension} + W_{normal} + W_{friction}$$

The left-hand side of the equation is now the change in the total mechanical energy of the box. The right-hand side of the equation is the sum of the work done by the non-conservative forces. It's clear now that the work done by the non-conservative forces on an object is what *changes* the total mechanical energy of an object. No potential energy can be associated with a non-conservative force because the work it does depends not only on the endpoints of the movement but also on the exact path taken.

In the next unit we will discuss in detail the calculation of the work done by non-conservative forces, especially friction, which is responsible for the change in the mechanical energy of an object.

MAIN POINTS

Potential Energy

The change in potential energy that is associated with a specific conservative force as an object moves between two locations is defined as negative the work done by that force between those two locations.

$$\Delta U = U_{\mathrm{B}} - U_{\mathrm{A}} \equiv -W_{\mathrm{A} \rightarrow \mathrm{B}}$$

A potential energy function can be defined for the object and the particular force by choosing a specific location as the zero of the function.

$$U = -W_{r \rightarrow r_o} + U_o$$

Conservation of Mechanical Energy

The mechanical energy of an object is defined to be the sum of its kinetic and potential energies.

$$E_{Mechanical} \equiv K + U$$

The work-kinetic energy theorem can be reformulated as a conservation law. Whenever the work done by non-conservative forces is zero, the mechanical energy of that object is conserved.

$$W_{Non-Conservative\ Forces} = 0 \quad \Rightarrow \quad \Delta E_{Mechanical} = 0$$

PROBLEMS

1. Pendulum: A mass $m = 5.3$ kg hangs on the end of a massless rope that has a length $L = 2.17$ m long. The pendulum is held horizontal and released from rest. (a) How fast is the mass moving at the bottom of its path? (b) What is the magnitude of the tension in the string at the bottom of the path? (c) If the maximum tension the string can withstand without breaking is $T_{max} = 438$ N, what is the maximum mass that can be used? (Assume that the mass is still released from the horizontal and swings down to its lowest point.) (d) A peg is placed 4/5 of the way down the pendulum's path so that when the mass falls to its vertical position it hits and wraps around the peg. As it wraps around the peg and attains its maximum height, it ends a distance of $3L/5$ below its starting point (or $2L/5$ from its lowest point). How fast is the mass moving at the top of its new path (directly above the peg)? (e) Using the original mass of $m = 5.3$ kg, what is the magnitude of the tension in the string at the top of the new path (directly above the peg)?

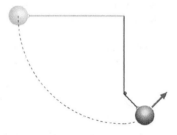

FIGURE 8.7 Problem 1 parts (d)-(e)

2. Loop the Loop: A mass $m = 75$ kg slides on a frictionless track that has a drop, followed by a loop-the-loop with radius $R = 18.2$ m and finally a flat straight section at the same height as the center of the loop (18.2 m off the ground). Since the mass would not make it around the loop if released from the height of the top of the loop (do you know why?) it must be released above the top of the loop-the-loop height. Assume the mass never leaves the smooth track at any point on its path. (a) What is the minimum speed at the top of the loop to make it around the loop-the-loop without falling off? (b) What height above the ground must the mass begin

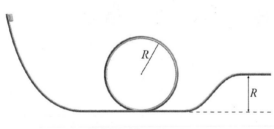

FIGURE 8.8 Problem 2

to make it around the loop-the-loop? (c) What is the speed of the mass at ground level? (d) What is the speed of the mass at the final flat level (18.2 m off the ground)? (e) A spring with spring constant $k = 17,200$ N/m is used on the flat surface to stop the mass. How far does the spring compress? (f) It turns out the engineers designing the loop-the-loop didn't really know physics–when they made the ride, the first drop was only as high as the top of the loop-the-loop. To account for the mistake, they decide to give the mass an initial velocity right at the beginning. How fast do they need to push the mass at the beginning (now at a height equal to the top of the loop-the-loop) to get the mass around the loop-the-loop without falling off the track? (g) Is the work done by the normal force on the mass (during the initial fall) *positive, zero*, or *negative*?

3. Block, Spring, and Incline (INTERACTIVE
EXAMPLE): A 5 kg block is placed near the top of a
frictionless ramp, which makes an angle of 30° to the
horizontal. At a distance $d = 1.3$ m away from the
block is an unstretched spring with $k = 3,000$ N/m. The
block slides down the ramp and compresses the spring.
Find the maximum compression of the spring.

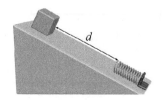

FIGURE 8.9 Problem 3

UNIT

9

WORK AND POTENTIAL ENERGY

9.1 Overview

This unit is concerned with two topics. We will first discuss the relationship between the real work done by kinetic friction on a deformable body and the calculation that we can perform using the work-kinetic energy theorem to determine the change in the mechanical energy due to kinetic friction. We will then determine how to describe a conservative force directly in terms of its potential energy function. We will use this understanding to develop a description of the equilibrium conditions for objects acted on by conservative forces.

9.2 Box Sliding Down a Ramp

Figure 9.1 shows a box sliding down a ramp. In Unit 6, we identified the forces acting on the box as shown in the figure and applied Newton's second law to determine that the

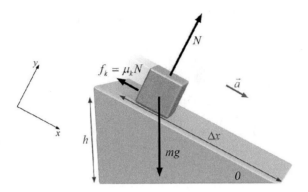

FIGURE 9.1 A box slides down a ramp. Three forces act on the box. Newton's second law can be used to determine the acceleration of the box.

acceleration of the box was proportional to the acceleration due to gravity, with the constant of proportionality being determined by the angle of the ramp and the coefficient of kinetic friction.

$$a = g(\sin\theta - \mu_k \cos\theta)$$

Since this acceleration is constant, we can apply our kinematics equations we derived for such a motion to determine the speed of the box at it reaches the end of the ramp, if it were released from rest.

$$v^2 = 2a\,\Delta x$$

$$v^2 = 2gh(1 - \mu_k \cot\theta)$$

We can also apply the work-kinetic energy theorem here, treating the box as a particle. In this case, the change in the mechanical energy of the box is equal to the work done by the non-conservative forces. The normal force does no work since it is perpendicular to the motion. Therefore, the change in mechanical energy of the box is just equal to the integral of the frictional force over the displacement.

$$\frac{1}{2}m_{box}v^2 - m_{box}gh = \int \vec{f}_k \cdot d\vec{x}$$

Writing the friction force in terms of the coefficient of kinetic friction and the normal force and applying Newton's second law to relate the normal force to the weight, we can obtain an expression for the final velocity.

$$\frac{1}{2}m_{box}v^2 - m_{box}gh = \mu_k N\,\Delta x = \mu_k(m_{box}g\cos\theta)\Delta x = \mu_k m_{box}g\cos\theta\left(\frac{h}{\sin\theta}\right)$$

$$v^2 = 2gh(1 - \mu_k \cot\theta)$$

This expression that we have obtained from the work-kinetic energy theorem is identical to that obtained by using Newton's second law as it must be! In the next section we will discuss the connection between the work done by kinetic friction when we consider the box to be a deformable body and the calculation we can make to determine correctly the change in the mechanical energy of such an object.

9.3   Work Done by Kinetic Friction

Figure 9.2 shows a box with an initial velocity $\vec{v}_o$ skidding across the floor, coming to rest a distance D from its initial position.

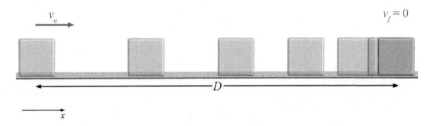

FIGURE 9.2 A box, with initial velocity v_o, skids across a floor and comes to rest a distance D from its initial position.

We can use the free-body diagram shown in Figure 9.3 to determine the acceleration of the box.

$$a = -\frac{\mu_k N}{m_{box}} = -\mu_k g$$

We can then use a constant acceleration kinematic equation to determine the distance D.

$$v_f^2 - v_o^2 = 2aD$$

$$D = \frac{v_o^2}{2\mu_k g}$$

We can also calculate this distance using the work-kinetic energy theorem by setting the change in kinetic energy equal to the **macroscopic work done by kinetic friction**.

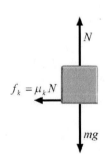

FIGURE 9.3 The free-body diagram for the sliding box in Figure 9.2.

$$-\frac{1}{2}m_{box}v_o^2 = -\mu_k m_{box}gD$$

We've used the words "macroscopic work" here to indicate that we are interested only in the macroscopic motion of the box. To obtain this motion, we have treated the box as a *single object*. At the microscopic level, the box does have deformable surfaces and a calculation of the microscopic work done by friction must account for the interactions at these surfaces. The good news is that we do not need to know anything about these microscopic details to correctly calculate the macroscopic motion of the box. In the next unit we will discuss systems of particles and introduce the concept of the *center of mass* that will justify this claim. Indeed, throughout this course we will be concerned only with *macroscopic mechanical energy*; this energy can be understood solely in terms of Newton's second law.

The underlying physics in this example that *cannot* be understood solely in terms of Newton's second law is the thermodynamics needed to understand why, as the box comes to rest, it actually gets hot. Namely, friction converts the macroscopic kinetic energy of the box into microscopic (thermal) energy of the molecules in the box and the floor. Understanding the details of this process is well beyond the scope of this course, yet we can get a qualitative picture of the mechanism involved by considering a simple model in which the atoms that make up the materials are thought of as little balls connected by springs. The frictional force then is modeled as the interactions of protrusions as the surfaces move past each other. The protrusions deform as they make contact, compressing and stretching some of the springs and causing a force opposing the relative motion. This is the force we call friction. Eventually the protrusions snap back past each other, causing vibrations of the balls that propagate to neighboring balls via the springs. These increased vibrations imply an increase in temperature. In other words, the deformation and release of the points of contact as the surfaces move past each other cause both the frictional force and the heating of the box.

9.4   Forces and Potential Energy

It may be useful at this point to summarize what we have learned so far. We can calculate the work done on an object by a force as it moves through some displacement by integrating the dot product of the force and the displacement. The total macroscopic work done on an object by all forces is equal to the change in the kinetic energy of the object.

For conservative forces such as gravity and springs, we defined a quantity called the change in the potential energy (ΔU) as negative the work done by that conservative force. The potential energy U for an object at a particular location was defined as the change in potential energy of the object between an arbitrary point that defines the zero of potential energy and the specified location.

We obtain the potential energy associated with a force by calculating the work done by that force.

$$\Delta U_{conservative\,F} = -W_{conservative\,F} = -\int \vec{F} \cdot d\vec{x}$$

Can we invert this process? That is, can we start with the potential energy of an object at a particular location and determine the force that is acting on the object at that location?

To answer this question, let's consider the one-dimensional case. If we look at the work done by the force as the object moves from x to $x + dx$, we see that the potential energy changes by an amount $dU = -F\,dx$. Consequently, we see that the force acting on the object at a given point is just equal to minus the derivative of the potential energy function at that same point.

$$F(x) = -\frac{dU(x)}{dx}$$

The force is a measure of how fast the potential energy is changing!

Of course, this result is not too surprising; if we find the potential energy from the force by evaluating an integral, then it's reasonable that we can find the force by performing the inverse operation, namely by taking the derivative of the potential energy. Indeed, this result can be generalized to more than one dimension with the use of the gradient operator ($\vec{F} = -\vec{\nabla}U$), but we will only deal with examples that are essentially one-dimensional here. In the next section, we will verify directly that differentiating the expressions we have derived for the potential energy associated with gravity and the spring force yield the familiar force expressions.

9.5   Examples: Force from Potential Energy

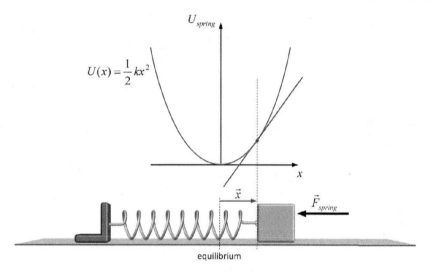

FIGURE 9.4 Defining the potential energy of a spring to be equal to zero at its equilibrium position results in the potential energy function being a parabola.

We will now consider a simple example to illustrate our result that the force acting on an object at a particular location is related to the spatial derivative of the potential energy of the object at that location. Figure 9.4 shows a mass attached to a spring. Defining x to be the displacement of the mass from its equilibrium position, we see that the potential energy function for the mass is a simple parabola.

The spatial derivative of the potential energy of the mass at point x is represented on the graph as the slope of the tangent to the curve at point x. *The steeper the slope, the bigger the magnitude of the force.* Taking the derivative of the potential function,

$$F(x) = -\frac{dU(x)}{dx} = -kx$$

we find that the force is proportional to the displacement from equilibrium, in agreement with our expression for the force law for springs.

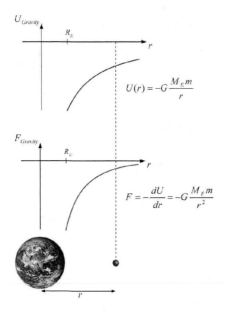

FIGURE 9.5 Differentiating the gravitational potential energy function recovers the familiar inverse square form of the Newton's universal gravitational force.

The same result also applies for the gravitational force as shown in Figure 9.5. Namely, we know that the potential energy of an object out in space some distance R from the center of Earth is inversely proportional to R. The slope of the tangent for this $1/R$ potential as a function of R tells us that the magnitude of the gravitational force is largest at the surface of the Earth and decreases as $1/R^2$ as we move farther away. Since this slope is always positive, the force will always be negative, indicating that the direction of the force always points toward the center of the Earth. This result is just what we expect for Newton's universal law of gravitation.

Note that shifting the potential energy function up or down in either example does *not* change the force since adding a constant to a function does not change its slope. This result affirms our understanding that we are always free to add a constant to the potential energy function without changing the underlying physics.

9.6 Equilibrium

So far we have used the word **equilibrium** to mean the position or orientation where the net force on an object is zero. For conservative forces, we can get an equivalent condition for equilibrium in terms of the potential energy. Namely, since we can express a conservative force in terms of the spatial derivative of its potential energy function, we see that the locations for equilibrium will be at points at which the slope of this potential energy function is zero!

For example, if we refer back to Figure 9.4, we see the potential energy function for a mass on a spring. The slope of the tangent to this curve is zero at only one point, namely, the point at which the potential energy itself is at its minimum.

Indeed, the slope of the tangent to *any* function will be zero at *any* minima or maxima of the function. Figure 9.6 shows a roller coaster track. Since the potential energy due to gravity for an object near the surface of the Earth is just equal to the product of the weight of the object and its height above a position defined to be zero, we see that the height of the track (measured from the zero of potential energy) is directly proportional to the gravitational potential energy of a car at that point. If a car is placed at rest at the minimum (B) or either of the two maxima (A) and (C), it will remain at rest since the net horizontal force on the car at each of these positions is zero. If the car at position (B) is given a small push in either direction it will return to the equilibrium position. If the car is given a small push in either direction at positions (A) or (C), however, it will roll down the track, moving farther away from the equilibrium position. We say that the equilibrium is *stable* at the minima and is *unstable* at the maxima.

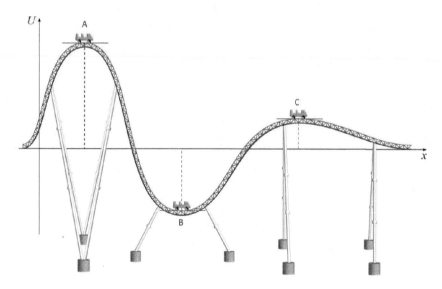

FIGURE 9.6 The points of equilibrium on a roller coaster track are those at which the slope (blue lines) of the track is zero.

To generalize this observation, we can say that an object which is initially at rest tends to move toward the configuration where its potential energy is minimized. Note that there is no new physics in this statement–this statement in terms of potential energy is equivalent to the statement that an object accelerates in the direction of the net force acting on it.

MAIN POINTS

Macroscopic Work Done by Friction

The work done by a kinetic friction force acting on a deformable object cannot be calculated without an understanding of the nature of the interactions at the surfaces.

However, the work-kinetic energy theorem can still be applied if we consider all forces acting at the center of mass of the object. The integral of the friction force through the full displacement of the object is called the macroscopic work done by friction. This macroscopic work does determine the change in the mechanical energy of the object.

$$W_{macroscopic\ friction} \equiv \int \vec{f}_k \cdot d\vec{x}$$

$$W_{macroscopic\ friction} = \Delta E$$

If friction is the only non-conservative force acting

Force from Potential Energy

For conservative forces, the force acting on any object at any point is equal to minus the spatial derivative of the potential energy function of the object at that point.

(1-D)

$$F(x) = -\frac{dU(x)}{dx}$$

(3-D)

$$\vec{F} = -\vec{\nabla}U$$

Equilibrium positions for a particle acted on by conservative forces are those locations where the slope of the potential energy function is zero.

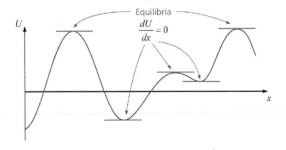

Equilibria

$$\frac{dU}{dx} = 0$$

PROBLEMS

1. Block on Incline: A mass $m = 12$ kg is pulled along a horizontal floor with no friction for a distance $d = 5$ m. Then the mass is pulled up an incline that makes an angle $\theta = 31°$ with the horizontal and has a coefficient of kinetic friction $\mu_k = 0.4$. The entire time the massless rope used to pull the block is pulled parallel to the incline at an angle of $\theta = 31°$ (thus on the incline it is parallel to the surface) and has a tension $T = 54$ N. (a) What is the work done by tension before the block goes up the incline? (b) What is the speed of the block right before it begins to travel up the incline? (c) What is the work done by friction after the block has traveled a distance of 2.2 m up the incline? (d) What is the work done by gravity after the block has traveled a distance of 2.2 m up the incline? (e) How far up the incline does the block travel before coming to rest as measured along the incline? (f) On the incline, is the net work done on the block is *positive*, *negative*, or *zero*?

2. Trip to the Moon: You plan to take a trip to the Moon. Since you do not have a traditional spaceship with rockets, you will need to leave the Earth with enough speed to make it to the Moon. Some information that will help during this problem:

$M_{Earth} = 5.97 \times 10^{24}$ kg
$R_{Earth} = 6.37 \times 10^{6}$ m
$M_{Moon} = 7.35 \times 10^{22}$ kg
$R_{Moon} = 1.737 \times 10^{6}$ m
$D_{Earth\ to\ Moon} = 3.844 \times 10^{8}$ m (center to center)
$G = 6.67428 \times 10^{-11}$ N-m^2/kg^2

FIGURE 9.7 Problem 2

(a) On your first attempt you leave the surface of the Earth at $v = 5,534$ m/s. How far from the center of the Earth will you get? (b) You consult a friend who calculates (correctly) the minimum speed needed as $v_{min} = 11,068$ m/s. If you leave the surface of the Earth at this speed, how fast will you be moving at the surface of the Moon? (c) Which of the following would change the minimum velocity needed to make it to the Moon: *the mass of the Earth, the radius of the Earth*, and/or *the mass of the spaceship*?

UNIT

10

CENTER OF MASS

10.1 Overview

This unit expands our study of mechanics from single particles to systems of particles. We will introduce the very important concept of the center of mass of a system of particles and determine the center of mass for both discrete and continuous mass distributions. We will use Newton's second law to obtain the equations of motion for the center of mass of a system of particles. We will also obtain a version of the work-kinetic energy theorem, called the *center-of-mass equation*, that can be applied to a system of particles.

10.2 Systems of Particles and the Center of Mass

So far we have only considered the motion of simple objects. We have not considered intentionally the motion of, for example, an object composed of two different sized balls connected to the ends of a rod. In the next few units we will develop the tools to understand the motion of more complicated systems of objects such as these. We will discover that their behavior can be understood by applying what we already know, and we will see that the equations describing their motion are remarkably similar to those we have already developed.

We start by introducing a new concept which will play a key role in what follows, namely that of the center of mass. Quite simply put, the **center of mass** of an object is just the average location of the mass that makes up the object. For a simple symmetric object like a ball or box of uniform density, we will see that the center of mass is just at the center of the object. For less simple shapes we will have to perform a calculation to determine the location of the center of mass.

The procedure we will adopt for finding the average position of all of the mass contained in some system of objects will be simply to take a mass-weighted average of the positions of the individual parts. Namely, we will define the location of the center of mass of a system of particles to be equal to the sum of the positions of the individual particles with each one weighted by its own fraction of the total mass of the system, as shown for a system of discrete masses in Figure 10.1.

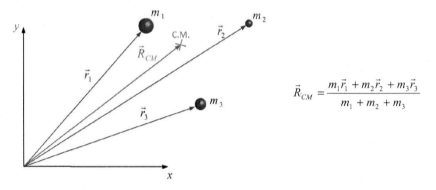

FIGURE 10.1 The definition of the center of mass for a system of three discrete masses.

In the next section we will determine the center of mass for a two-particle system. This concrete example will illustrate the general procedure and hopefully make clear why this definition makes sense.

10.3 Center of Mass for a Two-Body System

We'll start by considering an object made of only two point particles, labeled 1 and 2. We will assume that we know the masses of the two particles as well as their locations along the x-axis as shown in Figure 10.2.

Since the particles lie along the x-axis, the calculation of the position of their center of mass is straightforward.

$$\vec{R}_{CM} \equiv \frac{m_1\vec{r}_1 + m_2\vec{r}_2}{m_1 + m_2} \qquad \rightarrow \qquad x_{CM} \equiv \frac{m_1 x_1 + m_2 x_2}{m_1 + m_2}$$

FIGURE 10.2 Two masses located on the *x*-axis.

If the masses are equal, we see that the center of mass is located at the average value of x_1 and x_2. This position is halfway between them, which certainly makes sense. If, on the other hand, one mass is twice as big as the other, it will count for twice as much in the average, which means the center of mass will be closer to the heavier particle than the lighter one, which also seems reasonable.

So far we have considered just the one-dimensional case in which both particles lie along the *x*-axis. This procedure can be easily extended to more than one dimension, though, using vector addition. We start by writing the expression for the location of the center of mass in terms of the masses of the two particles and the vectors that locate each of the two particles.

$$\vec{R}_{CM} \equiv \frac{m_1\vec{r}_1 + m_2\vec{r}_2}{m_1 + m_2}$$

We can rewrite this formula so that the vector locating the center of mass is equal to the sum of two vectors, the displacement vector of one particle and another vector that is proportional to the difference in the displacement vectors of the two particles.

$$\vec{R}_{CM} \equiv \frac{m_1\vec{r}_1 + m_2\vec{r}_2}{m_1 + m_2} = \vec{r}_1 + \frac{m_2}{m_1 + m_2}\left(\vec{r}_2 - \vec{r}_1\right)$$

We can think of this equation as a map that tells us how to get to the center of mass: we first go to one of the objects and then we go a fraction of the way to the other object, where this fraction is determined by the masses, as shown in Figure 10.3. If the second object has the same mass as the first, we go half way. If the second object is heavier than the first, we go more than half way, and if the second is lighter than the first, we go less than halfway.

The beauty of this approach is that we can see that the location of the center of mass *does not* depend on our choice of the origin or the orientation of our coordinate system; the center of mass always lies at the same fixed point along the line connecting the two objects. We have just demonstrated an important result–namely, that the center of mass is a property of the system itself. It does not depend on the way we choose to look at the system. Indeed, we will show in a later unit that the center of mass of a rigid system is the same as its balance point!

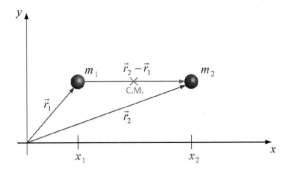

FIGURE 10.3 The location of the center of mass of a two-body system is located along the vector difference ($\vec{r}_2 - \vec{r}_1$) of the two displacements.

10.4   Center of Mass for Systems of More than Two Particles

We can extend our definition of the center of mass for systems containing more than two particles by simply summing up the mass-weighted displacement vectors for each particle.

$$\vec{R}_{CM} \equiv \frac{1}{M_{Total}} \sum_i m_i \vec{r}_i$$

It is usually easier to break this vector equation into components and evaluate each component separately. Just to make sure we know how this works, let's do an example involving eight equal mass particles located on the corners of a cube, as shown in Figure 10.4.

To find the x coordinate of the center of mass, we need to sum the x coordinates of each of the eight particles weighted by the ratio of the mass of each particle to the total mass of the system. In this example the particles all have the same mass and their x coordinates are either zero or L, so that the sum is easy to evaluate, and we find that the x coordinate of the center of mass is just equal to ½ L.

$$X_{CM} \equiv \frac{1}{M_{Total}} \sum_i m_i x_i = \frac{1}{8m} 4mL = \frac{L}{2}$$

We will get the same results for both the y and z coordinates, and we see that the center of mass is at the center of the box, as expected.

Now, suppose the cube was a solid, made up of millions of atoms rather than just eight particles on the corners. We suspect the answer would be the same, that the center of mass is still in the middle, but how can we prove this conjecture when actually performing the sum over millions of atoms seems difficult, if not impossible? Once again, calculus comes to the rescue!

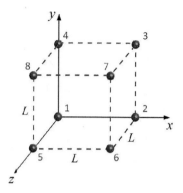

FIGURE 10.4 A system of eight equal mass particles located at the corners of a cube of side L.

Rather than calculating the product of position and mass for individual particles, we just integrate the position vector over all infinitesimal mass elements dm contained in the cube. We will do this calculation in the next section.

10.5   Center of Mass for Continuous Mass Distributions

Before we actually evaluate any integral, let's make sure we understand exactly how we adjust our definition of the center of mass when we are dealing with a continuous mass distribution. The big idea is that we have to replace a discrete sum by a continuous sum, an integral. In the discrete sum we evaluated the product of the mass of each part of the system and its position and then added them up. In the continuous sum we are doing the same thing, the only difference now is that we are dividing a continuous object up into an infinite number of tiny volume elements, each having a mass dm.

$$\vec{R}_{CM} \equiv \frac{1}{M_{Total}} \int \vec{r}\, dm$$

Since here we are integrating over the volume of a cube, a three-dimensional object, the integral itself must be evaluated in all three dimensions, x, y and z. To evaluate the mass element dm, we take the product of the volume of the element and the mass density (the mass per unit volume ρ) of the cube.

$$dm = \rho\, dV = \rho\, dxdydz$$

Our job now is to evaluate this triple integral. The first step is to break our equation for the center of mass vector $\vec{R}_{CM}$ into x, y and z components. We will start with the x equation and first determine the limits of integration. In each direction, the cube is located between the origin and a distance L from the origin. If we assume the mass density ρ is a constant, then it can be taken outside of the integral.

$$X_{CM} = \frac{1}{M_{Total}} \rho \int_0^L x\,dx \int_0^L dy \int_0^L dz$$

The resulting three-dimensional integral is equal to the product of three one-dimensional integrals, each of which is evaluated separately. Requiring the product of the mass density and the volume of the cube to be equal to the total mass of the cube, we obtain the expected result, that the coordinate of the center of mass of the cube is just equal to ½ L.

$$X_{CM} = \frac{1}{M_{Total}} \rho \left(\frac{1}{2} L^2 \right)(L)(L) = \frac{\rho L^3}{M_{Total}} \left(\frac{1}{2} L \right) = \frac{1}{2} L$$

The y and z component calculations are absolutely identical to the x component calculation, giving us the expected result: the center of mass of the box is at its center!

10.6   Center of Mass of a System of Objects

We now know how to find the center of mass of a collection of point particles as well as that of a continuous solid object. What happens when we want to find the center of mass of a collection of solid objects? Figure 10.5 shows two objects, labeled a and b. By definition, the center of mass of the system is found by integrating the position vector over all of the mass in the system. Since the system is made of two objects, the total integral is just the sum of two separate integrals, one for each object.

$$\vec{R}_{CM} = \frac{1}{M_{Total}} \left(\int_a \vec{r}\,dm + \int_b \vec{r}\,dm \right)$$

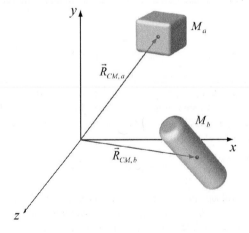

FIGURE 10.5 The center of mass of the two objects can be calculated simply by treating each object as a point particle having a mass equal to the total mass of the object.

If we multiply and divide each of these integrals by the mass of the object, we certainly haven't changed anything, but we can now see that the numerator just becomes the mass of each object times the position of its center of mass.

$$\vec{R}_{CM} = \frac{1}{M_{Total}}\left[M_a\left(\frac{\int_a \vec{r}\,dm}{M_a}\right) + M_b\left(\frac{\int_b \vec{r}\,dm}{M_b}\right)\right] = \frac{1}{M_{Total}}\left(M_a\vec{R}_{CM,a} + M_b\vec{R}_{CM,b}\right)$$

We have just arrived at a simple procedure for finding the center of mass of a system of solid objects. Namely, we just treat each object as a point particle with all of its mass located at its center of mass! That's all there is to it!

10.7   Dynamics of the Center of Mass

To this point we have defined the concept of the center of mass and have shown how to find it for any system of objects. With this knowledge in hand, we can finally do some physics.

We will start with our definition for the center of mass of a system of objects and take the derivative of this expression with respect to time.

$$\frac{d\vec{R}_{CM}}{dt} = \frac{1}{M_{Total}}\sum_i m_i \frac{d\vec{r}_i}{dt}$$

The left-hand side of the equation becomes the velocity of the center of mass, and the numerator on the right-hand side becomes the sum of the mass times velocity for each object in the system. We have already defined the product of the mass and velocity of an object as its momentum. Therefore, the numerator on the right-hand side is just equal to the total momentum of the system.

$$\vec{V}_{CM} = \frac{1}{M_{Total}}\sum_i m_i\vec{v}_i$$

We will now take another derivative with respect to time. The left-hand side of the equation becomes the acceleration of the center of mass, and the numerator on the right-hand side becomes the sum of the mass times acceleration for each object in the system.

$$\frac{d\vec{V}_{CM}}{dt} = \frac{1}{M_{Total}}\sum_i m_i \frac{d\vec{v}_i}{dt}$$

The product of the mass and acceleration of an object is just equal to the total force on that object.

$$\vec{A}_{CM} = \frac{1}{M_{Total}}\sum_i m_i\vec{a}_i = \frac{\sum_i \vec{F}_{Net,i}}{M_{Total}}$$

This sum of the total forces acting on all the objects in the system could get unwieldy if the number of objects gets large. The good news, though, is that we can simplify this sum significantly by realizing that any forces that act between two objects that are both in the system will *cancel* in this sum. Newton's third law requires that all such forces always come in pairs of equal magnitude and opposite direction!

$$\sum_i \vec{F}_{Net,i} = \sum_i \vec{F}_{External,i} + \sum_{i \neq j} \vec{F}_{ji} = \sum_i \vec{F}_{External,i}$$

Hence, the sum of all forces acting on all objects in the system just reduces to the sum of all forces acting on the system from the outside, or the total external force.

$$\vec{A}_{CM} = \frac{\sum_i \vec{F}_{External,i}}{M_{Total}} = \frac{\vec{F}_{Net,External}}{M_{Total}}$$

We have finally arrived at something that looks exactly like Newton's second law, but rather than applying just to point particles as before, this expression relates the total external force on the whole system to the acceleration of the center of mass of the system! Consequently, we can say that no matter how complicated a system of objects may be, the center of mass of the system behaves in the same simple way that a point particle does.

Indeed, in the next section, we will use this equation to obtain a generalization of the work-kinetic energy theorem for systems of particles.

10.8   The Center-of-Mass Equation

In Section 5 of Unit 7, we integrated Newton's second law to obtain the work-kinetic energy theorem for point particles: namely, that the change in kinetic energy of a particle is equal to the work done on that particle by the net force.

$$\Delta K = W_{Net}$$

We will now extend this result to systems of particles and start from our result from the last section. The acceleration of the center of mass of a system of particles is equal to the total external force acting on a system of particles divided by the total mass of the system.

$$\vec{A}_{CM} = \frac{\sum_i \vec{F}_{External,i}}{M_{Total}} = \frac{\vec{F}_{Net,External}}{M_{Total}}$$

This equation looks exactly like Newton's second law for a point particle that has the total mass of the system (M_{Total}) and is located at the center of mass of the system. Consequently, we can perform exactly the same derivation we made in Unit 7 to obtain the work-kinetic energy theorem for a point particle. The result here is an equation, often called the **center-of-mass equation**, which looks exactly like the work-kinetic energy theorem we derived for point particles.

$$\Delta\left(\frac{1}{2}M_{Total}V_{CM}^2\right) = \int \vec{F}_{Net,External} \cdot d\vec{l}_{CM}$$

The only differences lie in the subscripts. These subscripts *are important*, however. The change in kinetic energy term is calculated as if the system was a particle of mass M_{Total} and was moving with the velocity of the center of mass. We define the right-hand side of the equation as the **macroscopic work done by the net force**. This macroscopic work is calculated as if all forces were acting on this particle located at the center of mass.

We can now see why we did not need to worry about the microscopic work done by kinetic friction when we were calculating the motion of the box skidding to a stop in the last unit. If we consider the box to be a system of particles, we see that the change in kinetic energy of the box is exactly equal to the macroscopic work done by the net force, the kinetic friction force. The microscopic work done by the kinetic friction force at the interface of the surfaces of the box and the floor determines the additional thermal energy in the box and the floor, but does *not* determine the motion of the center of mass of the box.

10.9 Example: The Astronaut and the Wrench

We will end with a simple example to illustrate some of the concepts we have developed in this unit. Imagine you are an astronaut far out in space. You have just finished fixing a space telescope using a big wrench whose mass is one tenth that of yours. You realize you have no way to get back to your spaceship, which is 20 meters away from you, so you throw the wrench as hard as you can in a direction opposite from the spaceship, causing you to move in the opposite direction, toward the spaceship. When you finally reach the spaceship, how far away are you from the wrench?

The key concept needed to answer this question is that the acceleration of the center of mass of a system will be zero if the external force on the system is zero. In this case, we define the system to be you and the wrench, and the center of mass of the system is initially at rest a distance of 20 meters from your spaceship. Since there are no external forces acting on the system and the center of mass is initially at rest, the location of the center of mass of the system can never change! If we choose the initial location of the center of mass to be at $x = 0$, the center of mass will always be at $x = 0$.

The location of the center of mass of the system is determined from its definition.

$$X_{CM} = \frac{M_{astronaut}\, x_{astronaut} + M_{wrench}\, x_{wrench}}{M_{astronaut} + M_{wrench}} = 0$$

Multiplying both sides by the total mass, we obtain our result that the product of the position and mass of the wrench is always equal to minus the product of your position and mass.

$$M_{wrench}\, x_{wrench} = -M_{astronaut}\, x_{astronaut}$$

Since the mass of the wrench is 1/10 of your mass, the wrench will always be ten times as far away from the center of mass as you are, and it will always be on the opposite side of the center of mass from you.

$$x_{wrench} = -\frac{M_{astronaut}}{M_{wrench}} x_{astronaut}$$

When you are at the spaceship, 20 meters to the left of the center of mass, the wrench will be 200 meters to the right of the center of mass, which is 220 meters from the spaceship.

MAIN POINTS

Definition of Center of Mass

The center of mass of a system of objects is defined to be the mass-weighted average of its components.

(Discrete Particles)

$$\vec{R}_{CM} = \frac{1}{M_{Total}} \sum_{i=1}^{N} m_i \vec{r}_i$$

(Continuous Mass Distribution)

$$\vec{R}_{CM} = \frac{1}{M_{Total}} \int \vec{r} \, dm$$

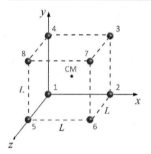

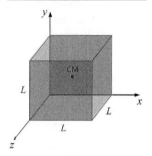

Equation of Motion for the Center of Mass

Applying Newton's second law to a system of particles, we obtain the equation of motion for the center of mass.

$$\vec{F}_{Net,External} = M_{Total} \vec{A}_{CM}$$

The Center-of-Mass Equation

Integrating the equations of motion for the center of mass, we obtain the center-of-mass equation that relates the change in the kinetic energy of the center of mass (*calculated as if the system were a particle having the total*

$$\Delta\left(\frac{1}{2} MV_{CM}^2\right) = \int \vec{F}_{Net,External} \cdot d\vec{l}_{CM}$$

mass of the system and moving with the velocity of the center of mass) to the macroscopic work done by the toal external force (*calculated as if all forces were acting at the center of mass*).

PROBLEMS

1. Man on a Boat: A man with mass $m_{man} = 60$ kg stands at the left end of a uniform boat with mass $m_{boat} = 161$ kg and a length $L = 3.1$ m. Let the origin of our coordinate system be the man's original location as shown in the drawing. Assume there is no friction or drag between the boat and water. (a) What is the location (i.e., the x coordinate) of the center of mass of the system? (b) If the man now walks to the right edge of the boat, what is the location of the center of mass of the system? (c) After walking to the right edge of the boat, how far has the man moved from his original location? (d) After the man walks to the right edge of the boat, what is the new location of the center of the boat? (e) The man walks to the very center of the boat. At what x coordinate does the man end up?

FIGURE 10.6 Problem 1

2. Playing Catch: A person with mass $m_L = 67$ kg stands at the left end of a uniform beam with mass $m_{beam} = 98$ kg and a length $L = 2.9$ m. Another person with mass $m_R = 70$ kg stands on the far right end of the beam and holds a medicine ball with mass $m_{ball} = 16$ kg (assume that the medicine ball is at the far right end of the beam as well). Let the origin of our coordinate system be the left end of the original position of the beam as shown in the drawing. Assume there is no friction between the beam and floor.

FIGURE 10.7 Problem 2

(a) What is the location of the center of mass of the system? (b) The medicine ball is thrown to the left end of the beam (and caught). What is the location of the center of mass now? (c) What is the new x position of the person at the left end of the beam? (How far did the beam move when the ball was thrown from person to person?) (d) To return the medicine ball to the other person, both people walk to the center of the beam. At what x position do they end up?

3. System of Particles: Four particles are in a 2-D plane with masses, x and y positions, and x and y velocities as given in the table shown. (a) What is the x position of the center of mass? (b) What is the y position of the center of mass? (c) What is the speed of the center of mass? (d) When a fifth mass is placed at the origin, what happens to the horizontal x location of the center of mass? Does it *move to the right, move to the left, not move,* or *it cannot be determined unless you know the mass?* (e) When a fifth

	m	x	y	v_x	v_y
1	7.8 kg	−2.6 m	−4.8 m	3.2 m/s	−3.9 m/s
2	9.0 kg	−3.4 m	3.6 m	−5.0 m/s	5.2 m/s
3	8.1 kg	4.7 m	−5.4 m	−6.3 m/s	2.0 m/s
4	9.3 kg	5.5 m	2.7 m	4.1 m/s	−3.3 m/s

mass is placed at the center of mass, what happens to the vertical y location of the center of mass: *it moves up, it moves down, it does not move,* or *it cannot be determined unless you know the mass?*

11

CONSERVATION OF MOMENTUM

11.1 Overview

This unit introduces the important concept of the conservation of momentum. Namely, the total momentum of a system of particles will be conserved whenever the sum of the external forces acting on the system is zero. We will apply this conservation law to collisions of particles and investigate sources of energy loss in these collisions.

We will also introduce a special reference frame, associated with a system of particles, called the center-of-mass frame, in which the total momentum of all the particles in the system is zero. The description of collisions is often simple in this frame.

11.2 Momentum Conservation

In the last unit we introduced the concept of the center of mass of a system of particles as the mass-weighted average of their positions. Taking the derivative of this expression with respect to time, the left-hand side of the equation becomes the velocity of the center of mass, and the numerator on the right-hand side becomes the sum of the mass times velocity for each object in the system.

$$\frac{d\vec{R}_{CM}}{dt} = \frac{1}{M_{Total}} \sum_i m_i \frac{d\vec{r}_i}{dt}$$

We have already defined the product of the mass and velocity of an object as its momentum. Therefore, the numerator on the right-hand side is just equal to the total momentum of the system. Multiplying both sides of the equation by the total mass of the system, we obtain the result that the total momentum of the system is equal to the product of the total mass and the velocity of the center of mass.

$$\vec{P}_{Total} = M_{Total}\vec{V}_{CM}$$

Differentiating once more with respect to time, we see that the time rate of change of the total momentum of the system is equal to the product of the total mass of the system and the acceleration of the center of mass.

$$\frac{d\vec{P}_{Total}}{dt} = M_{Total}\vec{A}_{CM}$$

We showed in the last unit that the product of the total mass and the acceleration of the center of mass is just equal to the total external force applied to the system. Therefore, we see that the time rate of change of the total momentum of the system is just equal to the total external force applied to the system.

$$\frac{d\vec{P}_{Total}}{dt} = \vec{F}_{Net,External}$$

This deceptively simple looking equation is extremely important, and we will spend several units exploring its meaning. $\vec{P}_{Total}$ is the total momentum vector of the system and is equal to the vector sum of the momenta of all of the parts of the system. Likewise, the total external force is the vector sum of all external forces acting on all parts of the system.

Note that when there are no external forces acting on the system, the time rate of change of the total momentum is zero. In other words, if the total external force is zero, then the total momentum of the system does not change in time. In this case, we say that the *momentum of the system is conserved*. We will now work out a couple of simple examples that illustrate momentum conservation.

11.3 Momentum Example: Astronaut and Wrench

We'll start by revisiting the problem we ended with in the last unit–that of an astronaut throwing a wrench. Since the astronaut and the wrench are both initially at rest, the initial momentum of the system is zero, and since there are no external forces acting on the system of the astronaut and the wrench, the total momentum of the system is conserved and will therefore *always* be zero. The total momentum of the system has two contributions–one from the astronaut and one from the wrench–and the vector sum of these is zero.

We see that in order for the total momentum to be zero, the astronaut must move in the opposite direction of the wrench with a speed fixed by the ratio of the masses.

$$\vec{P}_{astronaut} = -\vec{P}_{wrench} \qquad \rightarrow \qquad \vec{v}_{astronaut} = -\frac{m_{wrench}}{m_{astronaut}} \vec{v}_{wrench}$$

For example, if the mass of the astronaut is ten times as big as the mass of the wrench, the speed of the wrench will be ten times the speed of the astronaut. This requirement ensures both that the magnitudes of the momentum of the wrench and the astronaut are the same, and also that the center of mass of the system does not move since the distance the wrench moves in any given time interval will be ten times that moved by the astronaut.

Let's now examine our momentum equation a bit more carefully. Suppose the total external force is zero in some direction but not in others–what can we say about the momentum of the system?

$$\frac{d\vec{P}_{Total}}{dt} = \vec{F}_{Net,External}$$

Since our equation is a vector equation, we know that the only component of momentum which will be conserved will be the one that lies along the direction in which the total external force is zero.

11.4 Example: An Inelastic Collision

We will now consider an interesting class of problems that can be addressed using this conservation of momentum principle. Namely, we will look at collisions between particles. We'll start with the example shown in Figure 11.1. A box of mass m_1 slides with velocity v_1 along a horizontal frictionless floor and collides with a second box of mass m_2 which is initially at rest. After the collision the boxes stick together and move with a final velocity v_f. Our job is to determine this final velocity.

In this problem the system we are interested in is made up of the two boxes. Since the floor is horizontal and frictionless, the total external force on the system in the horizontal direction is zero. Therefore, the total momentum of the two box system is conserved. That is, this total momentum will be the same before and after the collision.

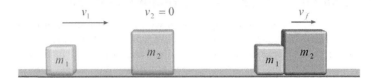

FIGURE 11.1 An inelastic collision: Box 1 moves with speed v_1 and collides with box 2 that is initially at rest. The two boxes stick together and move off with speed v_f. The momentum of the system of two boxes is conserved in this collision which allows us to determine the final speed v_f.

$$\vec{P}_{initial} = \vec{P}_{final}$$

At this point you might be wondering about the forces that will act between the boxes during the actual collision itself–won't these forces, which will definitely have components in the horizontal direction, change the momentum of the system? The answer is no, they will not, and the reason is simple: The forces between the boxes are not external forces; these forces are internal forces, being exerted by the boxes, the objects that make up the system. In other words, the force by box 1 on box 2 will definitely change the momentum of box 2 and the force by box 2 on box 1 will definitely change the momentum of box 1, but the total momentum of the two boxes will not change since these forces are equal and opposite by Newton's third law! For this reason we never actually have to worry about what happens during the instant when the boxes collide. We can just focus on the total momentum before and after.

In this example the initial momentum of the system in the horizontal direction is due entirely to box 1. The final momentum of the system is due to *both* boxes. Since the initial and final momentum of the system has to be the same,

$$m_1 v_1 = (m_1 + m_2) v_f$$

we can solve for the final velocity of boxes 1 and 2 in terms of the initial velocity of box 1.

$$v_f = \frac{m_1}{(m_1 + m_2)} v_1$$

11.5   Energy in Collisions

We see that momentum is conserved in this collision, but what happens to the total kinetic energy of the system? Is it conserved also? The answer to this question depends on what we call the kinetic energy of the system. Certainly, if there are no external forces acting on the system, then there is no macroscopic work done on the system and the kinetic energy of the system, defined as one-half the total mass times the square of the velocity of the center of mass cannot change either. However, this kinetic energy of the center of mass is *not* equal to the sum of the kinetic energies of the objects making up the system. We will demonstrate this claim now as we explicitly calculate the sum of the kinetic energies of boxes 1 and 2 before and after the collision.

Before the collision, the sum of the kinetic energies of the boxes is just equal to the initial kinetic energy of box 1.

$$K_{initial} = \frac{1}{2}m_1 v_1^2$$

After the collision, the sum of the kinetic energies of the boxes is equal to the final kinetic energy of the object composed of the two boxes stuck together.

$$K_{final} = \frac{1}{2}(m_1 + m_2)v_f^2$$

In the last section, we used the conservation of momentum to determine the final velocity of the boxes in terms of the initial velocity of box 1. Therefore, we can determine the final kinetic energy in terms of the initial kinetic energy.

$$K_{final} = \frac{1}{2}(m_1 + m_2)\left(\frac{m_1}{m_1 + m_2}v_1\right)^2 = K_{initial}\left(\frac{m_1}{m_1 + m_2}\right)$$

We see here that the final energy is smaller than the initial kinetic energy by exactly the same factor that related the final and initial velocities. In other words, the kinetic energy of the system, defined as the sum of the kinetic energies of the boxes, was *not* conserved in the collision. We call this kind of a collision an **inelastic collision**. In the next section, we will look at how this energy is lost.

11.6   Energy Loss in Collisions

We saw in the last section that the kinetic energy of the boxes after the collision was less than the kinetic energy of the boxes before the collision. How can we understand this loss of energy? Where did the energy go?

To understand this loss of energy, we need to look at the collision in more detail. Let's first focus our attention on box 1. We can define box 1 to be our system and apply the center-of-mass equation to determine that the change in the kinetic energy of the box is equal to the macroscopic work done on the box during the collision.

$$\Delta K_1 = \int \vec{F}_{21} \cdot \vec{dl}_{1_{CM}}$$

What force is responsible for this work? Clearly, the force that box 2 exerts on box 1 during the collision must be responsible for this work. This work is done during the time of the collision and it may be hard to visualize since the idealized diagram we have drawn seems to suggest that the boxes themselves are not crushed or deformed during the collision.

To get a better feeling for what is going on, just consider what happens to two cars after a collision. The obvious deformation of the cars as a result of the collision shows where the energy was lost during the collision. This energy loss can be understood in terms of the work done during this collision by the force that one car exerts on the other times the distance that the front of the other car was deformed.

Returning to our example of the sliding boxes, we can see that if we actually wanted the boxes to stick together we would have to provide some mechanism for non-conservative work to be done during the collision. Perhaps we could put a bit of putty on the surface of one of the boxes that could be compressed during the collision. The details of the nature of the internal forces acting during the collision can influence the amount of energy lost in a collision, but as long as there are no external forces acting, then we can be sure that the total momentum of the system will be conserved!

11.7   Center-of-Mass Reference Frame

We will now return to the concept of the center of mass since we will find that it can play a useful role in collisions as well. We have already derived the important relationship between the total momentum of a system and the velocity of its center of mass.

$$\vec{P}_{Total} = M_{Total}\vec{V}_{CM}$$

If we know that the total momentum does not change in time, for example, then it must be true that the velocity of the center of mass also does not change in time!

Recall the example of the astronaut throwing the wrench. We determined that the velocity of the center of mass of the system (astronaut + wrench) was constant and, in fact, equal to zero. The total momentum of the system was zero implying that the momentum of the wrench was exactly equal and opposite to the momentum of the astronaut. The reference frame in which we presented this example is called the **center-of mass-reference frame**, since the velocity of the center of mass is zero in this frame.

What about the more general case when the center of mass is moving with some constant velocity? We already know how to compare measurements in different reference frames. We learned in Unit 3 that if the velocity of an object is known in reference frame A, and reference frame A is moving relative to reference frame B with a constant velocity, then the velocity of the object in reference frame B is just equal to the vector sum of these velocities.

$$\vec{v}_{object,B} = \vec{v}_{object,A} + \vec{v}_{A,B}$$

Therefore, once we determine the velocity of the center of mass in the given frame, we can always transform the problem to the center-of-mass frame, if doing so makes the problem easier to solve. We will do such an example in the next section.

11.8   Example: Center-of-Mass Reference Frame

Suppose an asteroid is moving with a constant velocity of 4 km/s in the $+x$ direction as observed by a spaceship. An explosive device inside the asteroid suddenly blows it into two chunks, one having twice the mass of the other as shown in Figure 11.2. In the reference frame of the asteroid, the lighter chunk moves in the $+y$ direction with a speed of 6 km/s. What is the speed of the heavier chunk of the asteroid as measured by someone on the spaceship?

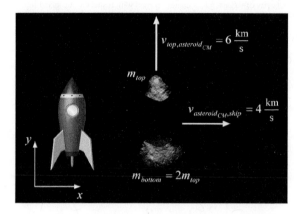

FIGURE 11.2 An asteroid moving in the x direction suddenly explodes into two pieces. Conservation of momentum is most conveniently applied in the asteroid center of mass to determine the speed of the heavier chunk in the spaceship frame.

The total momentum is *always zero* in the center-of-mass reference frame. The center-of-mass frame for the two chunks is clearly the frame in which the asteroid was at rest before it exploded. Since the total momentum is zero in this frame, the momentum of the two chunks after the explosion must be equal and opposite. If the lighter chunk has a velocity of 6 km/s upward then the bigger chunk must be moving downward and must have half the speed of the lighter chunk since it has twice the mass.

The velocity of any object in the reference frame of the spaceship is equal to the velocity of that object in the center-of-mass reference frame plus the velocity of the center of mass in the reference frame of the ship.

$$\vec{v}_{object,ship} = \vec{v}_{object,CM} + \vec{v}_{CM,ship}$$

For the big chuck of asteroid, the velocity relative to the center of mass is 3 km/s in the $-y$ direction and the velocity of the center of mass relative to the ship is 4 km/s in the $+x$ direction. We can add these vectors using the Pythagorean theorem to find that the speed of the big chunk is 5 km/s in the spaceship frame as shown in Figure 11.3.

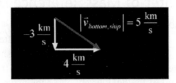

FIGURE 11.3 To find the velocity of the chunk with respect to the spaceship, we take the vector sum of the velocity of the chunk with respect to the asteroid's center of mass and the velocity of the asteroid's center of mass with respect to the spaceship.

MAIN POINTS

Conservation of Momentum

If the sum of the external forces acting on any system of particles is zero, then the total momentum of the system, defined as the vector sum of the momenta of the individual particles, is conserved.

$$\text{When } \vec{F}_{Net,External} = 0 \Rightarrow \vec{P}_{Total} \equiv \sum_i \vec{p}_i = \text{Constant}$$

Forces in a Collision

Internal forces determine the amount of energy lost in a collision.

If only internal forces act during a collision, the total momentum of the system will be conserved.

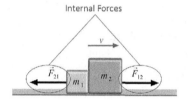

Internal Forces

Center-of-Mass Reference Frame

The center-of-mass reference frame is defined to be that frame in which the total momentum of all particles in the system is zero. In other words, it is the frame in which the center of mass of the system is at rest.

Center-of-Mass Reference

$$\vec{P}_{Total} = M_{Total}\vec{V}_{CM} = 0$$

Conservation of momentum calculations are often simplified in the center-of-mass frame.

PROBLEMS

1. Car-Truck Collision: A car with mass $m_C = 494$ kg is moving east with a speed of $v_C = 22$ m/s and collides with a truck with mass $m_T = 1,209$ kg that is moving south with a speed of $v_T = 14$ m/s . The two collide and lock together after the collision. (a) What is the magnitude of the initial momentum of the car? (b) What is the magnitude of the initial momentum of the truck? (c) What is the angle that the car-truck combination travels after the collision? Give your answer as an angle south of east. (d) What is the magnitude of the momentum of the car-truck combination immediately after the collision? (e) What is the speed of the car-truck combination immediately after the collision? (f) Compare the initial and final kinetic energy of the total system before and after the collision: $K_i = K_f$, $K_i > K_f$, or $K_i < K_f$?

2. Two Train Cars: A train car with mass $m_1 = 618$ kg is moving to the east with a speed of $v_1 = 7.7$ m/s and collides with a second train car. The two cars latch together during the collision and then move off to the east at $v_f = 4.7$ m/s. (a) What is the initial momentum of the first train car? (b) What is the mass of the second train car? (c) What is the change in kinetic energy of the two-car system during the collision? (d) The same two cars are involved in a second collision. The first car is again moving to the east with a speed of $v_1 = 7.7$ m/s and collides with the second train car that is now moving to the west with a velocity $v_2 = -5.7$ m/s before the collision. The two cars latch together at impact. What is the final velocity of the two-car system? (A positive velocity means the two train cars move to the east, and a negative velocity means the two train cars move to the west.) (e) How does the magnitude of the momentum of train car 1 compare before and after the collision: $p_{1\,initial} = p_{1\,final}$, $p_{1\,initial} > p_{1\,final}$, or $p_{1\,initial} < p_{1\,final}$?

3. Explosion: An object with a total mass of $m_{total} = 17.5$ kg is sitting at rest when it explodes into three pieces. One piece with mass $m_1 = 4.9$ kg moves up and to the left at an angle of $\theta_1 = 22°$ above the $-x$ axis with a speed of $v_1 = 27.5$ m/s. A second piece with mass $m_2 = 5.2$ kg moves down and to the right an angle of $\theta_2 = 27°$ to the right of the $-y$ axis at a speed of $v_2 = 20.8$ m/s. (a) What is the magnitude of the final momentum of the system (all three pieces)? (b) What is the mass of the third piece? (c) What is the x component of the velocity of the third piece? (d) What is the y component of the velocity of the third piece? (e) What is the magnitude of the velocity of the center of mass of the pieces after the collision? (f) Calculate the increase in kinetic energy of the pieces during the explosion.

FIGURE 11.4 Problem 3

4. Bullet, Block, and Spring (INTERACTIVE EXAMPLE): A bullet with a mass of 20 g and a speed of 100 m/s collides with a wooden block of mass 2 kg. The wooden block is initially at rest on a smooth table and is connected to a spring with a spring constant of $k = 800$ N/m. The other end of the spring is attached to an immovable wall. What is the maximum compression of the spring? (*Note*: You may assume that the spring is massless and that the collision between the bullet and the wooden block is completely inelastic.)

FIGURE 11.5 Problem 4

5. Collision with Friction (INTERACTIVE EXAMPLE): A block of mass $M_1 = 3.5$ kg moves with velocity $v_1 = 6.3$ m/s on a frictionless surface. It collides with a block of mass $M_2 = 1.7$ kg which is initially stationary. The blocks stick together and encounter a rough surface. The blocks eventually come to a stop after traveling a distance $d = 1.85$ m. What is the coefficient of kinetic friction on the rough surface?

UNIT

12

ELASTIC COLLISIONS

12.1 Overview

In this unit, our focus will be on elastic collisions, namely those collisions in which the only forces that act during the collision are conservative forces. In these collisions, the sum of the kinetic energies of the objects is conserved. We will find that the description of these collisions is significantly simplified in the center-of-mass frame of the colliding objects. In particular, we will discover that, in this frame, the speed of each object after the collision is the same as its speed before the collision.

12.2 Elastic Collisions

In the last unit, we discussed the important topic of momentum conservation. In particular, we found that when the sum of the external forces acting on a system of particles is zero, then the total momentum of the system, defined as the vector sum of the individual momenta, will be conserved. We also determined that the kinetic energy of the system, defined to be the sum of the individual kinetic energies, is not necessarily conserved in collisions. Whether or not this energy is conserved is determined by the details of the forces that the components of the system exert on each other. In the last unit, our focus was on inelastic collisions, those collisions in which the kinetic energy of the system was not conserved. In particular non-conservative work was done by the forces that the individual objects exerted on each other during the collision.

In this unit, we will look at examples in which the only forces that act during the collision are conservative forces. In this case, the total kinetic energy of the system is conserved. We call these collisions, **elastic collisions**. As an example, consider the collision we discussed in the last unit with one modification–instead of having the boxes stick together, we'll put a spring on one of the boxes as shown in Figure 12.1. The spring will compress during the collision, storing potential energy, and when it relaxes back to its original length it will turn

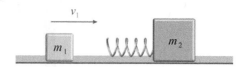

FIGURE 12.1 An elastic collision: Box 1 moves with speed v_1 and collides with box 2 that is initially at rest. A spring is connected to box 2 that is initially at rest. A spring is connected to box 2 which is compressed during the collision and then extends to send the two boxes in opposite directions. The mechanical energy of the two boxes and spring system is conserved

this stored potential energy back into kinetic energy. In this way, no mechanical energy is lost during the collision so that the final kinetic energy of the system will be the same as its initial kinetic energy.

12.3   One-Dimensional Elastic Collisions

We will start with the example from the last section. Knowing that neither the momentum nor the kinetic energy of the system will change during this collision allows us to write down two independent equations that relate the initial and final velocities of the boxes.

$$m_1 v_{1,i} + m_2 v_{2,i} = m_1 v_{1,f} + m_2 v_{2,f} \qquad \text{(Conservation of momentum)}$$

$$\frac{1}{2} m_1 v_{1,i}^2 + \frac{1}{2} m_2 v_{2,i}^2 = \frac{1}{2} m_1 v_{1,f}^2 + \frac{1}{2} m_2 v_{2,f}^2 \qquad \text{(Conservation of kinetic energy)}$$

These two equations contain six variables (the initial and final velocities of box 1, the initial and final velocities of box 2, and the masses of the two boxes). Therefore, if we know any four of these quantities, these two equations will allow us to solve for the other two. For example, if we know the masses and the initial velocities of both boxes then we can solve these two equations for the final velocities of both boxes.

There is a complication, however, that will make the actual solution of these equations tedious, at best. For example, if we solve the momentum equation for the velocity of box 2 after the collision in terms of the velocity of box 1 after the collision, and plug the result into the energy equation, we get a pretty messy quadratic equation that, with some effort, can certainly be solved.

$$\frac{1}{2}m_1v_{1,i}^{\,2}+\frac{1}{2}m_2v_{2,i}^{\,2}=\frac{1}{2}m_1v_{1,f}^{\,2}+\frac{1}{2}m_2\left(v_{2,i}+\frac{m_1}{m_2}(v_{1,i}+v_{1,f})\right)^2$$

There is, however, a better way. Physics can rescue us from this tedious mathematical chore! Namely, if we solve this problem in the center-of-mass frame, we can avoid solving any quadratic equations. We can then determine the final velocities in the initial frame by simply transforming the velocities in the center-of-mass frame back into the initial frame. We will perform this calculation in the next section.

12.4   The Center-of-Mass Frame

Figure 12.2 shows the collision as viewed in the center-of-mass frame. We have labeled the velocities in this frame with an asterisk (*).

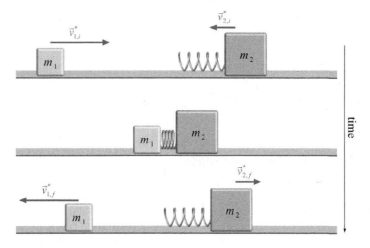

FIGURE 12.2 The elastic collision as viewed in the center-of-mass frame.

We know the total momentum is conserved in *any* inertial reference frame. What distinguishes the center-of-mass frame, though, and what simplifies the calculation, is the additional constraint that the *total momentum is always zero in the center-of-mass frame.* Consequently, the single momentum conservation equation has become two equations, one for the initial state and one for the final state.

$$m_1v_{1,i}^{*}-m_2v_{2,i}^{*}=0$$

$$-m_1v_{1,f}^{*}+m_2v_{2,f}^{*}=0$$

Note that we have chosen to use the speed variables (the magnitude of the velocities) in these equations. Therefore, we have explicitly inserted the minus signs to indicate the direction of the velocities. To see how this constraint simplifies the problem, we will multiply and divide each term in the energy equation by the appropriate mass. The result of this operation is that each term now is proportional to the square of an individual momentum.

$$\frac{1}{2m_1}(m_1 v_{1,i}^*)^2 + \frac{1}{2m_2}(m_2 v_{2,i}^*)^2 = \frac{1}{2m_1}(m_1 v_{1,f}^*)^2 + \frac{1}{2m_2}(m_2 v_{2,f}^*)^2$$

We can now use the momentum equations to write each side of the energy equation in terms of the square of the momentum of just one of the particles.

$$\left(\frac{1}{2m_1} + \frac{1}{2m_2}\right)(m_1 v_{1,i}^*)^2 = \left(\frac{1}{2m_1} + \frac{1}{2m_2}\right)(m_1 v_{1,f}^*)^2$$

In fact, we see that the magnitude of the momentum of each of the objects individually is now also the same before and after the collision, although the direction of each one has changed. In other words, when the collision is viewed in this reference frame, the *speed* of each object is the *same* before and after the collision.

$$v_{1,i}^* = v_{1,f}^* \qquad\qquad\qquad v_{2,i}^* = v_{2,f}^*$$

This result is very important; it provides us with a very simple strategy to solve any elastic collision problem.

12.5 Example: Center of Mass

We will now work out an example that demonstrates the use of the center-of-mass frame in elastic collisions. In the collision that was shown in Figure 12.1, we will assume $m_1 = 2$ kg, $v_1 = 5$ m/s, and $m_2 = 3$ kg. The boxes collide elastically and both move along the axis defined by the initial velocity vector (call it the x-axis). Our job is to determine the final velocities of both boxes in this reference frame, which we will call the lab frame.

Our first task is to transform this problem to the center-of-mass system. In order to make this transformation, we need to know the velocity of the center of mass in the lab frame. In Unit 10 we determined that this velocity was just equal to the vector sum of the individual velocities, weighted by the fraction of the total mass each particle carries.

$$\vec{V}_{CM,lab} = \frac{m_1 \vec{v}_1 + m_2 \vec{v}_2}{m_1 + m_2} = \frac{m_1}{m_1 + m_2}\vec{v}_1$$

Plugging in the values for the masses and the initial velocities, we find that the center of mass is moving at 2 m/s in the +x direction.

We can now use this value for the velocity of the center of mass to determine the initial velocities of the boxes as viewed in the center-of-mass frame. We know that the velocity of an object in the center-of-mass frame is equal to the velocity of the object in the lab frame plus the velocity of the lab frame in the center-of-mass frame.

$$\vec{v}^*_{object} = \vec{v}_{object.lab} + \vec{V}_{lab,CM}$$

We know the velocity of the lab in the center-of-mass frame must just be equal to minus the velocity of the center of mass in the lab frame. We can now find the initial velocities of both boxes in the center-of-mass frame by simply adding numbers that we now know. Namely, we find the initial velocity of box 1 in the center-of-mass frame is equal to 5 m/s – 2 m/s = 3 m/s in the positive x direction, and the initial velocity of box 2 in the center-of-mass frame is equal to 0 m/s – 2 m/s = –2 m/s in the x direction. From the last section we know that the final speeds in the center-of-mass frame are equal to initial speeds in that frame. Therefore, we know the final velocity of box 1 is equal to 3 m/s in the *negative x* direction, and the final velocity of box 2 is equal to 2 m/s in the *positive x* direction.

$$\vec{v}^*_{1,f} = -3 \text{ m/s } \hat{i} \qquad\qquad \vec{v}^*_{2,f} = +2 \text{ m/s } \hat{i}$$

Our final step is to transform these results to the lab frame. We can make this transformation by simply adding the velocity of each object in the center-of-mass frame to the velocity of the center of mass in the lab frame. When we make these additions, we see that after the collision, box 1 moves with speed 1 m/s in the negative x direction and box 2 moves with speed 4 m/s in the $+x$ direction. You can verify, using these values, that both momentum and energy are indeed conserved in this collision!

If we now replace the masses and initial velocities in this problem by variables and follow the identical procedure, we arrive at the general expressions for the final velocities of any two objects undergoing an elastic collision under the assumption that the second object is at rest to begin with and that all motion is in one dimension.

$$\vec{v}_{1,f} = \vec{v}_{1,i} \frac{m_1 - m_2}{m_1 + m_2} \qquad\qquad \vec{v}_{2,f} = \vec{v}_{1,i} \frac{2m_1}{m_1 + m_2}$$

We can learn a couple of interesting things from these equations. First, if the masses are the same, we find that final velocity of the first object is zero, and the final velocity of the second object is just equal to the initial velocity of the first. In other words, the objects trade roles!

Second, we see that the final velocity of the first object changes sign if m_2 is greater than m_1. In other words, if the first object is lighter than the second object, it will bounce back. If, on the other hand, the first object is heavier than the second object, it will continue in its initial direction with a reduced speed.

12.6   Elastic Collisions in Two Dimensions

The last example assumed that the motion of the colliding objects was constrained to one dimension. We found that knowing the masses and the initial velocities of both objects was enough to completely determine the final velocities. We will now extend this analysis to two dimensions by considering the collision of objects on a frictionless horizontal surface without any constraint that the motion is along a single axis. An actual example of such a situation might be billiard balls colliding on a pool table or pucks colliding on an air-hockey table.

We will start by making the simplification that the mass of both objects is the same and that one of the objects is initially at rest. Even in this restricted case, we can see that the final velocity vectors cannot be determined from a knowledge of the initial velocity vectors. Indeed, the final directions depend on the orientation of the objects when they collide. If the collision is head-on, the final velocities will be along the x-axis just like in the previous example, but if the collision is not head-on, as illustrated in Figure 12.3, the final velocities can have y components as well.

FIGURE 12.3 An elastic collision between two equal mass balls. If the centers of the balls are not aligned, the collision becomes two-dimensional; the final velocities can develop y components.

What can we say about the final velocities in these cases? Just as in the previous example, we get the simplest view of the collision from the center-of-mass reference frame as shown in Figure 12.4. Prior to the collision we see the objects approaching each other head on, and afterward we see them leaving the collision point back to back with exactly the same speed they had before the collision. The one parameter that is not determined from the conservation of momentum and energy is the angle between the initial and final velocities of one of the objects. Indeed, this angle can vary all the way from 180° in the case of a head-on collision, to 0° in the case where the objects simply miss each other.

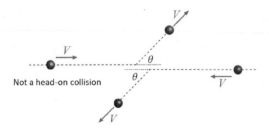

FIGURE 12.4 An elastic collision viewed from the center-of-mass frame. The angle θ is *not* determined from conservation of momentum and energy.

MAIN POINTS

Elastic Collisions

If the only forces acting during a collision are conservative forces, then the kinetic energy of the system, defined to be the sum of the kinetic energies of the colliding objects, is conserved. Such collisions are called elastic collisions.

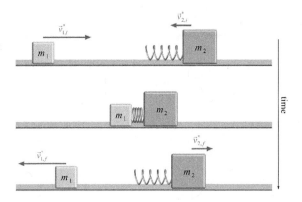

$$\sum K_i = \sum K_f$$

Center-of-Mass Frame

Elastic collisions are most simply described in the center-of-mass reference frame of the colliding objects.

Center-of-Mass Reference Frame

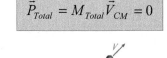

$$\vec{P}_{Total} = M_{Total}\vec{V}_{CM} = 0$$

The collision may cause objects to be deflected through some angle in the frame, but their speeds will always remain the same.

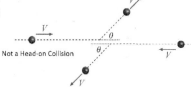

Not a Head-on Collision

PROBLEMS

1. Bumper Cars: A bumper car with mass $m_1 = 120$ kg is moving to the right with a velocity of $v_1 = 4.3$ m/s. A second bumper car with mass $m_2 = 83$ kg is moving to the left with a velocity of $v_2 = -3.9$ m/s. The two cars have an elastic collision. Assume the surface is frictionless. (a) What is the velocity of the center of mass of the system? (b) What is the initial velocity of car 1 in the center-of-mass reference frame? (c) What is the final velocity of car 1 in the center-of-mass reference frame? (d) What is the final velocity of car 1 in the ground (original) reference frame? (e) What is the final velocity of car 2 in the ground (original) reference frame? (f) In a new (inelastic) collision, the same two bumper cars with the same initial velocities now latch together as they collide. What is the final speed of the two bumper cars after the collision? (g) Compare the loss in energy in the two collisions: $|\Delta K_{elastic}| = |\Delta K_{inelastic}|$, $|\Delta K_{elastic}| > |\Delta K_{inelastic}|$, or $|\Delta K_{elastic}| < |\Delta K_{inelastic}|$.

2. Billiard Balls: A white billiard ball with mass $m_w = 1.65$ kg is moving directly to the right with a speed of $v = 3.22$ m/s and collides elastically with a black billiard ball with the same mass $m_b = 1.65$ kg that is initially at rest. The two collide elastically and the white ball ends up moving at an angle above the horizontal of $\theta_w = 69°$ and the black ball ends up moving at an angle below the horizontal of $\theta_b = 21°$. (a) What is the final speed of the white ball? (b) What is the final speed of the black ball? (c) What is the magnitude of the final total momentum of the system? (d) What is the final total energy of the system?

Before   After

FIGURE 12.5 Problem 2

13

COLLISIONS, IMPULSE AND REFERENCE FRAMES

13.1 Overview

In this unit we will conclude our discussion of collisions and look at the energy of a system of particles in more detail. In particular, we will start by developing a useful relation between relative velocities that must hold in an elastic collision. We will then look at the details of the collision process and introduce the concept of the impulse that describes the change in momentum of one of the objects in a collision. Finally, we will investigate the kinetic energy of a system of particles and find that the total kinetic energy can be expressed as the sum of the kinetic energy of the center of mass and the kinetic energy of the particles relative to the center of mass.

13.2   Relative Speed in Elastic Collisions

In the last unit we discovered that the description of collisions is often simplified when viewed in the center-of-mass reference frame. In particular, we showed that the speed of an object before and after an elastic collision is the same when viewed in this frame even though its direction will be changed as is shown in Figure 13.1. We will now use this result to obtain a relation between relative speeds in a collision that will hold in *all* reference frames.

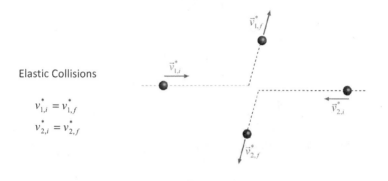

Elastic Collisions

$$v^*_{1,i} = v^*_{1,f}$$

$$v^*_{2,i} = v^*_{2,f}$$

Center-of-Mass Frame

FIGURE 13.1 An elastic collision as viewed in the center-of-mass frame. In this frame, the speeds of each particle do not change.

In particular, since the speed of an object before and after an elastic collision is the same if viewed in the center-of-mass frame, then it is also true that the relative speed of the two objects is the same before and after the collision in this frame.

$$\left| \vec{v}^*_{2,i} - \vec{v}^*_{1,i} \right| = \left| \vec{v}^*_{2,f} - \vec{v}^*_{1,f} \right|$$

That is, the rate at which two objects approach each other before an elastic collision is the same as the rate at which they separate afterward.

We can use this result to identify elastic collisions in any inertial reference frame. Namely, the relative velocity of two objects at a given time, that is, the difference in the velocity vectors of the objects must be the same in all inertial reference frames.

This claim follows from the fact that to transform both velocity vectors to a different inertial frame, we simply add the same vector (the relative velocity vector for the two frames) to each initial velocity vector. This relative velocity vector then cancels when we take the difference of the velocities of the objects.

$$\vec{v}_{2,B} - \vec{v}_{1,B} = \left(\vec{v}_{2,A} + \vec{v}_{A,B} \right) - \left(\vec{v}_{1,A} + \vec{v}_{A,B} \right) = \vec{v}_{2,A} - \vec{v}_{1,A}$$

If the relative *velocity* of two objects at a given time is the same in all inertial reference frames, then the relative *speed* of the two objects must also be the same in all inertial reference frames. Since we have just shown that the relative speed of the two objects in an elastic collision is the same before and after the collision in the center-of-mass frame, then it follows that the relative speed of the two objects in an elastic collision is the same before and after in any inertial reference frame!

Indeed, if we look back to the one-dimensional example in Section 5 of the last unit, we see that the relative speeds of the two objects, that is, the difference in the magnitudes of their velocities, is equal to 5 m/s both before *and* after the collision, in both the center of mass *and* the lab reference frames!

13.3   Elastic Collision Examples

We just showed that, in an elastic collision between two objects, the rate at which the objects approach each other before the collision is the same as the rate at which they separate after the collision and that this statement is true in all inertial reference frames!

$$\left| \vec{v}_{2,i} - \vec{v}_{1,i} \right| = \left| \vec{v}_{2,f} - \vec{v}_{1,f} \right|$$

For example, suppose we throw a ball against the wall of a building. If the wall is hard and solid and the ball is made of good hard rubber then the collision will be almost elastic, and we expect the speed of the ball to be about the same before and after it bounces off the wall.

Suppose we now consider a bowling ball, moving with speed V colliding head-on with a ping-pong ball that is initially at rest. If we assume the collision to be elastic and the motion to be constrained to one dimension, what will be the final velocities of the balls?

How do we go about solving this problem? The one thing we do know is that if the collision is elastic, the speed of the ping-pong ball relative to the bowling ball must be the same after the collision as it was before the collision. Before the collision, the speed of the ping-pong ball relative to the bowling ball was just equal to V. Therefore, the speed of the ping-pong ball relative to the bowling ball after the collision must also be equal to V.

We can obtain an approximate solution by assuming the velocity of the bowling ball will not change much during the collision since it is much heavier than the ping-pong ball. In this approximation, we expect the final speed of the ping-pong ball to be about twice the initial speed of the bowling ball.

As a check, we can look at the exact solution which we obtained in Section 5 of the last unit.

$$\vec{v}_{1,f} = \vec{v}_{1,i} \frac{m_1 - m_2}{m_1 + m_2}$$

$$\vec{v}_{2,f} = \vec{v}_{1,i} \frac{2m_1}{m_1 + m_2}$$

We see that in the limit where $m_1 \gg m_2$, we recover our approximate solution and that the speed of the ping-pong ball after the collision is about twice the speed of the bowling ball.

13.4 Forces During Collisions

In applying conservation of momentum to collisions between two objects, we have been concerned only with the velocities of the objects before and after the collision. We now want to investigate exactly what Newton's laws can tell us about the details of the collision process itself.

We start with the differential form of Newton's second law which relates the total force on an object to the time rate of change of its momentum.

$$\vec{F}_{Net} = \frac{d\vec{p}}{dt}$$

We can rewrite this expression to determine that the change in the momentum of an object during a small time dt is just equal to the total force acting on the object multiplied by this time interval. If we now integrate this expression over the time of the collision itself, we see that the total change in the momentum of the object during the collision is equal to the integral of the total force acting on that object during this time.

$$\int_{t_1}^{t_2} \vec{F}_{Net} \, dt = \int d\vec{p} = \vec{p}(t_2) - \vec{p}(t_1) \equiv \Delta\vec{p}$$

This integral is usually called the **impulse** delivered by the force.

We can use this result to define the average force acting on the object during the collision to be equal to the change in the momentum of the object divided by the duration of the collision.

$$\Delta\vec{p} = \int_{t_1}^{t_2} \vec{F}_{Net} \, dt \equiv \vec{F}_{avg} \, \Delta t$$

This result simply reflects the differential form of Newton's second law that we used to get started.

13.5 Impulse Examples

We have just determined that the change in momentum of an object during a collision is equal to the product of the average force acting on that object and the time over which it acts. Therefore, we can achieve the same change in momentum by having a large force acting for a short time as having a small force acting for a long time.

Figure 13.2 depicts an example to illustrate this observation. A ball with a mass of 1 kg is released from rest from an initial height of 1 meter above the floor. It bounces back to half its original height. If we assume the ball is in contact with the floor for a time of 10 ms, what is the average force on the ball during the collision?

To determine the average force acting during the collision, we need to first determine the change in the momentum of the ball. We can use the conservation of energy during the ball's initial free fall to determine its speed just before it hits the floor, and we find that it is proportional to the square root of the height from which it was released.

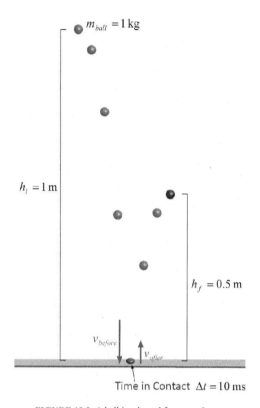

FIGURE 13.2 A ball is released from rest from a height of 1 m and rebounds to a height of 0.5 m. Assuming the ball was in contact with the floor for 10 ms, what was the average force exerted by the floor on the ball?

$$m_{ball}gh_i = \frac{1}{2}m_{ball}v_{before}^2$$

$$v_{before} = \sqrt{2gh_i}$$

Putting in the numbers, we obtain a speed of 4.43 m/s. We can also use energy conservation to determine that for the ball to rebound to a height of 0.5 m, it must have had a speed of 3.13 m/s immediately after it left the floor.

$$v_{after} = \sqrt{2gh_f}$$

The change in the momentum of the ball during the collision is therefore equal to 7.56 kg·m/s , since the initial direction is downward and the final direction is upward. We can now determine the average force acting during the collision by dividing this change in momentum by the duration of the collision to obtain the value of 756 N.

$$\Delta p = m_{ball}\Delta v_{ball}$$

Suppose we were to repeat the exact same experiment with a harder ball that flexes less and consequently spends less time in contact with the floor. If, for example, the time of the collision is reduced by a factor of ten, the average force on the ball must be increased by the same factor of ten to keep the change in momentum the same. In other words, the average force on the ball during such a collision would be 7,560 N.

13.6 Energy of a System of Particles

We have seen that often the simplest description of collisions occurs in a reference frame in which the center of mass of the colliding objects is at rest. We will now extend this approach to the discussion of the kinetic energy of a system of particles.

Consider a simple system made up of two point particles of mass m_1 and m_2 connected by a massless rod. If we throw this object in our laboratory reference frame we know it will tumble in some complicated way, but we also know that the center of mass will move in a very simple way, namely that the center of mass will behave as though it were a point particle having the total mass of the object as shown in Figure 13.3.

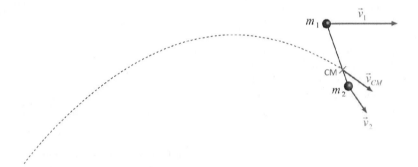

FIGURE 13.3 An object consisting of balls connected by a massless rod is in free fall. Although the motion of the individual balls is complicated, the center of mass of the system must follow the parabolic trajectory of any object in free fall.

At any instant the kinetic energy of the system is equal to the sum of the kinetic energies of the two particles.

$$K_{system,lab} = \sum_i \frac{1}{2}m_i\left(\vec{v}_i \cdot \vec{v}_i\right)$$

We can express the velocity of an object in the lab frame as the vector sum of the velocity of the object in the center-of-mass reference frame plus the velocity of the center of mass in the lab reference frame,

$$\vec{v}_i = \vec{v}_i^* + \vec{v}_{CM}$$

so that the kinetic energy of the system can be rewritten as

$$K_{system,lab} = \sum_i \frac{1}{2}m_i(\vec{v}_i \cdot \vec{v}_i) = \sum_i \frac{1}{2}m_i(\vec{v}_i^* + \vec{v}_{CM}) \cdot (\vec{v}_i^* + \vec{v}_{CM})$$

$$K_{system,lab} = \sum_i \frac{1}{2}m_i v_i^{*2} + \left(\frac{1}{2}\sum_i m_i\right)v_{CM}^2 + \left(\sum_i m_i\vec{v}_i^*\right) \cdot \vec{v}_{CM}$$

When we make this sum, we see that the total energy of the system as viewed in the lab frame can be written as the sum of just two terms:

$$K_{system,lab} = \sum_i \frac{1}{2}m_i v_i^{*2} + \frac{1}{2}M_{Total}v_{CM}^2 + \vec{P}_{Total,CM} \cdot \vec{v}_{CM}$$

$$K_{system,lab} = \sum_i \frac{1}{2}m_i v_i^{*2} + \frac{1}{2}M_{Total}v_{CM}^2$$

The first term is the sum of the kinetic energies of the objects as viewed in the center-of-mass reference frame,

$$K_{REL} \equiv \sum_i \frac{1}{2}m_i v_i^{*2}$$

and the second term is the kinetic energy of the center of mass as viewed in the lab reference frame,

$$K_{CM} \equiv \frac{1}{2}M_{Total}v_{CM}^2$$

The term that vanishes involves the total momentum in the center-of-mass reference frame, which by definition is always zero ($\vec{P}_{Total,CM} \equiv 0$).

The result we have just derived is completely general. The total kinetic energy of any system of objects as viewed by any observer is simply equal to the total kinetic energy of the objects as viewed in the center-of-mass reference frame, often called the relative kinetic energy (K_{REL}), plus the total kinetic energy of the center of mass in the observer's reference frame, often called the center-of-mass kinetic energy (K_{CM}).

$$K_{system,lab} = K_{REL} + K_{CM}$$

This result has two profound implications. First, the total kinetic energy of a system of particles will, in general, have two distinct components. This result will become central to our discussion of rotations in the next unit. Second, we see that the kinetic energy of a system of particles *does* depend on the reference frame of the observer. In other words, the relative kinetic energy will be the same for all observers, but the center-of-mass kinetic energy will be different for different observers since it will depend on the speed of the center of mass in the frame of the observer.

MAIN POINTS

Elastic Collisions: Relative Speeds

The rate at which two objects approach each other before an elastic collision is equal to the rate at which they separate afterward.

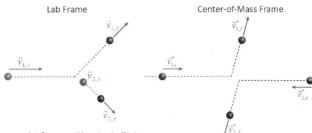

Lab Frame   Center-of-Mass Frame

Relative Speeds Before and After an Elastic Collision

$$\left| \vec{v}_{2,f} - \vec{v}_{1,f} \right| = \left| \vec{v}_{2,i} - \vec{v}_{1,i} \right| = \left| \vec{v}_{2,i}^{\,*} - \vec{v}_{1,i}^{\,*} \right| = \left| \vec{v}_{2,f}^{\,*} - \vec{v}_{1,f}^{\,*} \right|$$

Impulse

The impulse is defined as the integral of the force over the time of the collision.

Integrating Newton's second law over time, we find that the impulse is equal to the change in momentum during the collision.

Impulse $\displaystyle\int_{t_1}^{t_2} \vec{F}_{Net}\, dt = \Delta \vec{p}$

Kinetic Energy of a System of Particles

The kinetic energy of a system of particles, defined as the sum of the kinetic energies of the particles in the system, is equal to the kinetic energy of the particles relative to the center of mass, a term that is the same in all reference frames, plus the energy of the center of mass, a term that does depend on the reference frame of the observer.

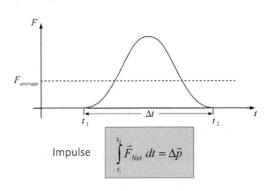

$$K_{system,lab} \equiv \sum_i \frac{1}{2} m_i v_i^2 = \sum_i \frac{1}{2} m_i v_i^{*2} + \frac{1}{2} M_{Total} v_{CM}^2$$

PROBLEMS

1. Ball Hits Wall: A racquet ball with mass $m = 0.253$ kg is moving toward the wall at $v = 13$ m/s and at an angle of $\theta = 25°$ with respect to the horizontal. The ball makes a perfectly elastic collision with the solid, frictionless wall and rebounds at the same angle with respect to the horizontal. The ball is in contact with the wall for $t = 0.065$ s. (a) What is the magnitude of the initial momentum of the racquet ball? (b) What is the magnitude of the change in momentum of the racquet ball? (c) What is the magnitude of the average force the wall exerts on the racquet ball? (d) The racquet ball is moving straight toward the wall at a velocity of $v_i = 13$ m/s. The ball makes an inelastic collision with the solid wall and leaves the wall in the opposite direction at $v_f = -8.5$ m/s. The ball exerts the same average force on the ball as before. What is the magnitude of the change in momentum of the racquet ball? (e) What is the time the ball is in contact with the wall? What is the change in kinetic energy of the racquet ball?

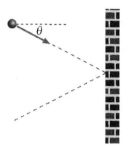

FIGURE 13.4 Problem 1

UNIT

14

ROTATIONAL KINEMATICS AND MOMENT OF INERTIA

14.1 Overview

In this unit we will introduce rotational motion. In particular, we will introduce the angular kinematic variables that are used to describe the motion and will relate them to our usual one-dimensional kinematic variables. We will also define the moment of inertia, the parameter in rotational motion that is analogous to the mass in translational motion. We will evaluate the moment of inertia for a collection of discrete particles as well as for symmetric solid objects.

14.2 Rotational Kinematics

Until now our studies of dynamics have been restricted to the linear motion of objects described in a Cartesian coordinate system (x, y, and z). In our recent discussions of systems of particles, though, we have discovered that the motion can be described as having two components: (1) the motion *of* the center of mass and (2) the motion *relative to* the center of mass. As an illustration of the motion relative to the center of mass, we will look at the rotation of an object about an axis through the center of mass. Our first step is to develop a coordinate system in which these rotations can be described naturally.

Figure 14.1 shows a disk rotating about an axis through its center. The orientation of the disk at any time can be described by a single parameter–the angle θ through which the disk has rotated relative to its initial orientation. We call this angle the **angular displacement**. The time rate of change of the angular displacement is called the **angular velocity**, ω, and the time rate of change of the angular velocity is called the **angular acceleration**, α.

FIGURE 14.1 The rotation of a disk about an axis through its center is described by an angular displacement θ.

$$\omega \equiv \frac{d\theta}{dt}$$

$$\alpha \equiv \frac{d\omega}{dt}$$

These equations look strikingly similar to those we used to describe one-dimensional kinematics.

$$v \equiv \frac{dx}{dt} \qquad\qquad a \equiv \frac{dv}{dt}$$

The reason for this similarity is simply that this rotational motion can be described by a single angular displacement, θ, just as linear motion can be described by a single spatial displacement, x.

If we now consider the special case of a constant angular acceleration α, we can derive the equations for ω and θ for this motion by integrating the defining equations. The resulting equations for ω and θ are absolutely identical in form to those describing one-dimensional motion at constant acceleration with the substitutions α for a, ω for v, and θ for x.

$$\theta = \theta_o + \omega_o t + \frac{1}{2}\alpha t^2$$

$$\omega = \omega_o + \alpha t$$

$$\omega^2 - \omega_o^2 = 2\alpha(\theta - \theta_o)$$

14.3   Relating Linear and Rotational Parameters

We can now make another useful connection between rotation and one-dimensional kinematics by obtaining the relationships between the angular and linear kinematic parameters used to describe the motion of a point that is a fixed distance R from the rotational axis.

In the case of one-dimensional motion along the x-axis, we needed to specify which direction we choose to be positive so that the signs of displacement, velocity, and acceleration have meaning. In exactly the same way, we need to specify which direction of rotation we choose to be positive so that the signs of angular displacement, angular velocity, and angular acceleration have meaning.

Although we are always free to choose either direction of rotation to be positive, it is customary to pick the counterclockwise as the positive direction. With this choice, the angular displacement θ agrees with that usually used in trigonometry. Since all points on the disk are rotating together, we can determine the linear displacement, speed, and acceleration of any point on the disk in terms of the corresponding angular parameters. Figure 14.2 shows a disk turned through an angular displacement θ. We can see that a point located a distance R from the rotation axis moves through a distance s along a circular path of radius R. This distance s is determined from geometry to be just equal to the product of the radius and the angular displacement, measured in radians.

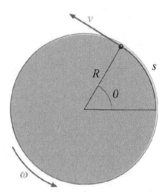

FIGURE 14.2 A disk rotates with angular velocity ω through an angular displacement θ.

$$s = R\theta$$

Taking the derivative of this linear distance with respect to time, we find a simple relationship between the speed of the point and the angular velocity of the disk.

$$\frac{ds}{dt} = R\frac{d\theta}{dt}$$

$$v = R\omega$$

Taking the derivative of this linear speed with respect to time, we find a simple relationship between the tangential acceleration of the point along its circular path and the angular acceleration of the disk.

$$\frac{dv}{dt} = R\frac{d\omega}{dt}$$

$$a = R\alpha$$

14.4   Kinetic Energy in Rotations

We will now expand our discussion of rotations by considering the motion of a rigid object made up of a set of point particles connected by massless rods as shown in Figure 4.3. This object rotates about the fixed axis with a constant angular velocity ω. We will assume we know the masses of each particle (m_i) and the distances of each particle from the axis of rotation (r_i).

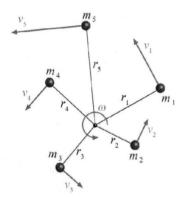

FIGURE 14.3   A rigid object consisting of five point particles connected by massless rods rotates with angular velocity ω.

The total kinetic energy of this object is defined to be the sum of the kinetic energies of each of its parts. We found in the last section that the speed of a rotating object relative to the axis of rotation is just the product of its angular velocity and its distance from the axis. Therefore, we can rewrite the expression for the total kinetic energy of the object in terms of its angular velocity.

$$K_{system} = \sum \frac{1}{2}m_i(r_i\omega)^2$$

Since this angular velocity is a constant, we can take it, and the common factor of one-half, outside the sum.

$$K_{system} = \frac{1}{2}\left(\sum m_i r_i^2\right)\omega^2$$

We will define the remaining sum, namely, the sum of the product of each mass with the square of its distance from the axis, to be the **moment of inertia** of the object about this axis and will denote it with the symbol I.

$$I \equiv \sum m_i r_i^2$$

. Note that the resulting expression for the kinetic energy of this object has the same form as the kinetic energy of a point particle.

$$K_{system} = \frac{1}{2} I \omega^2 .$$

We have just replaced the velocity by the angular velocity and mass by the moment of inertia.

In other words, in the same way that the mass of an object tells us how its kinetic energy is related to the square of its velocity, the moment of inertia of a rotating object tells us how its kinetic energy is related to the square of its angular velocity. In a sense, the moment of inertia plays the same role in rotational motion that the mass plays in the simpler motions we have studied up to this point. As we learn more about rotations, we will see this conceptual connection between mass and the moment of inertia appear again and again.

14.5   Moment of Inertia

We will spend the remainder of this unit exploring the properties of the moment of inertia in more detail. The most obvious difference between mass and the moment of inertia is that the moment of inertial depends not just on the total mass but also on exactly where that mass is located. Indeed, the moment of inertia even depends on our choice of the rotation axis since we are measuring all distances relative to this axis.

We will now do a simple example to illustrate these points. Figure 14.4 shows an object made up of four point particles of equal mass M arranged in a square of side $2L$ centered on the origin. The x- and y-axes are as shown and the z-axis points out of the page. We will first calculate the moment of inertia of this object if it is rotated around the x-axis. Since the distance of each mass from the x-axis is just equal to L, the sum we need to make to determine the moment of inertia is simple.

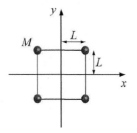

$$I_x \equiv \sum m_i r_i^2 = 4ML^2$$

Clearly, if we were to calculate the moment of inertia for rotations around the y-axis, we would find it to be identical to the moment of inertia for rotations about the x-axis.

FIGURE 14.4 A rigid object consists of four point particles, each of mass M, located at the corners of a square of side $2L$.

What about rotations around the z-axis? The distance of each particle from the z-axis is clearly larger than L. Indeed, using the Pythagorean theorem, we see that the square of the distance to each mass is twice as big as before.

$$I_z \equiv \sum m_i r_i^2 = 8ML^2$$

Therefore, the moment of inertia for rotations about the z-axis is twice as big as the moment of inertia for rotations about the x- or y-axes.

We've seen from this example exactly how to calculate the moment of inertia of an object made up of discrete point particles about any axis. We've learned that the moment of inertia does depend on the choice of the rotation axis. In the next section, we will generalize this calculation to the case of a continuous solid object, rather than a small collection of points.

14.6   Moment of Inertia of a Solid Object

Figure 14.5 shows a thin rod of mass M and length L centered along the x-axis. How do we go about calculating its moment of inertia for rotations about the z-axis? Once again, as we have done so often, we will need to replace the discrete sum we used in the last section with an appropriate integral.

We will call the mass per unit length of the rod λ. Each infinitesimal piece of the rod has a length dx and a mass equal to the product of this length and the mass per unit length.

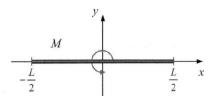

$$dm = \lambda \, dx$$

The contribution to the moment of inertia from such an infinitesimal piece of the rod located a distance x from the origin is just equal to the product of its mass and the square of its distance from the axis.

FIGURE 14.5 A thin rod of mass M and length L is aligned with the x-axis.

$$dI_z = dm \, x^2$$

To find the total moment of inertia, we integrate over the length of the rod.

$$I_z = \int_{-L/2}^{+L/2} x^2 \lambda \, dx = \lambda \int_{-L/2}^{+L/2} x^2 \, dx$$

The mass density can be taken outside the integral, and we are left with the integral of x^2 which is just $x^3 / 3$. Evaluating this expression between the limits, we obtain

$$I_z = \lambda \left[\frac{1}{3} x^3 \right]_{-L/2}^{L/2} = \lambda \left(\frac{1}{12} L^3 \right)$$

We can replace the mass density by the total mass divided by the length of the rod to obtain an expression that is proportional to the product of the total mass and the square of the length of the rod, as expected.

$$I_z = \frac{1}{12} ML^2$$

We know the moment of inertia depends on the choice of the axis. Suppose we want to calculate the moment of inertia about an axis that is parallel to the z-axis but passes through the end of the rod rather than its middle. How does the calculation change? To determine the moment of inertia we just do the integral again, this time shifting the location of the rod to the right so that its left end is at the origin.

$$I_z = \lambda \int_0^L x^2 \, dx = \frac{1}{3} \lambda L^3 = \frac{1}{3} ML^2$$

Evaluating the integral, we see that the moment of inertia about the end of the rod is four times as big as the moment of inertia about its center. This result is reasonable, since more of the mass is farther from the axis when it is located at the end of the rod rather than at its center.

14.7   Example: The Moment of Inertia of a Solid Cylinder

We will now do one more example that will illustrate some general features of moments of inertia of solid objects. Figure 14.6 shows a solid cylinder of mass M and radius R. The axis of the cylinder coincides with the z-axis, and its end surfaces are located at $z = 0$ and $z = L$.

Since this object has cylindrical symmetry, our integral will be simplified if we use cylindrical coordinates (namely, r, ϕ, and z) rather than Cartesian coordinates. Figure 14.7 shows the volume element illustrated as the product of dr, dz, and $r \, d\phi$. To integrate over the mass of the cylinder, we use the mass element dm, which is just equal to the product of this volume element and ρ, the mass per unit volume of the cylinder.

$$dm = \rho r \, dr \, dz \, d\phi$$

To evaluate the moment of inertia about the axis of symmetry, the z-axis, we just need to integrate $r^2 dm$ over the entire cylinder.

$$I_z = \int r^2 dm = \iiint \rho r^3 \, dr \, dz \, d\phi = \rho \int_0^L dz \int_0^{2\pi} d\phi \int_0^R r^3 \, dr$$

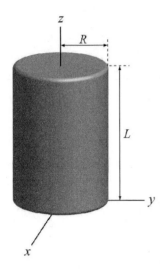

FIGURE 14.6 A solid cylinder of radius R, length L, and mass M.

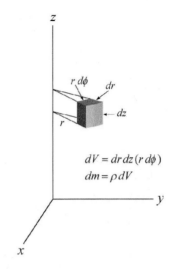

$$dV = dr\,dz\,(r\,d\phi)$$
$$dm = \rho\,dV$$

FIGURE 14.7 A volume element in cylindrical coordinates for the cylinder in Figure 14.6.

The z and ϕ integrals are trivial, being equal to $2\pi L$. We are left then with just the integral of r^3. Evaluating this integral and simplifying, we see that the moment of inertia of the cylinder about its axis is equal to $\frac{1}{2}MR^2$.

$$I_z = \frac{M}{\pi R^2 L}(2\pi)(L)\left(\frac{R^4}{4}\right) = \frac{1}{2}MR^2$$

Note that this result does not depend explicitly on the length of the cylinder, only on its mass and radius. For example, if we were to cut the cylinder in half through a plane perpendicular to the z-axis, we would have two cylinders, each with half the mass of the original cylinder. Therefore, the moment of inertia of each new cylinder is just half of the moment of inertia of the original cylinder.

$$I_{Total} = \sum_i I_i$$

Consequently, we see that if a system is made of two or more parts, and we know the moment of inertia of each part about some axis, then the total moment of inertia about that axis is just the sum of the moments of inertia of the parts. This result may seem somewhat trivial, but it will prove useful later.

14.8 Moment of Inertia for Solid Objects

We have just determined that the moment of inertia of a solid cylinder about its axis is proportional to the product of its mass and the square of its radius. We can determine the moment of inertia of any other object with cylindrical or spherical symmetry in exactly the same way. We will *always* find this same result, that the moment of inertia will be proportional to the product of its mass and the square of its radius:

$$I_{Total} \propto MR^2$$

The constant of proportionality will be different, of course, for each different shape. The larger this constant, the more mass is located far from the axis. For example, it is easy to see that for a cylindrical shell, this constant of proportionality is just equal to one, since all of the mass is located at the radius of the shell.

Consider a solid sphere and a solid cylinder. We expect the constant of proportionality for the sphere to be smaller than that for the cylinder since more of its mass is concentrated near the axis. When we do the calculation, we see that this constant for the sphere is two-fifths which is indeed smaller than the factor of one-half we calculated for the solid cylinder. Similarly, we expect that the constant for a spherical shell is smaller than that for a cylindrical shell.

$$I_{solid\ cylinder} = \frac{1}{2} MR^2 \qquad\qquad I_{cylindrical\ shell} = MR^2$$

$$I_{solid\ sphere} = \frac{2}{5} MR^2 \qquad\qquad I_{spherical\ shell} = \frac{2}{3} MR^2$$

MAIN POINTS

Rotational Kinematics

Rotational motion is described in terms of the (1) angular displacement θ, (2) the angular velocity ω, and (3) the angular acceleration α.

$$\omega \equiv \frac{d\theta}{dt}$$

$$\alpha \equiv \frac{d\omega}{dt}$$

The displacement, velocity, and acceleration of any point that is rotating is proportional to the corresponding angular parameter.

$$s = R\theta$$

$$v = R\omega$$

$$a = R\alpha$$

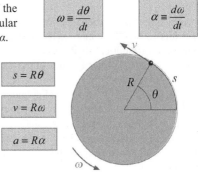

Moment of Inertia & Kinetic Energy

The moment of·inertia of a system of particles about an axis is defined to be the sum of the product of the mass and the square of the distance from the axis for all parts of the system.

Discrete Distributions

$$I \equiv \sum m_i r_i^2$$

Continuous Distributions

$$I = \int r^2 \, dm$$

Rotational Kinetic Energy

The kinetic energy of a system of particles is equal to one-half the product of the square of the angular velocity and the moment of inertia about the axis of rotation.

$$K_{system} = \frac{1}{2} I\omega^2$$

Moment of Inertia of Cylinders and Spheres

The moment of inertia for any cylindrical or spherical object is proportional to the product of the square of the radius of the object and the total mass of the object.

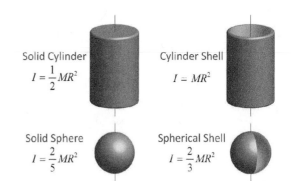

Solid Cylinder

$$I = \frac{1}{2} MR^2$$

Cylinder Shell

$$I = MR^2$$

Solid Sphere

$$I = \frac{2}{5} MR^2$$

Spherical Shell

$$I = \frac{2}{3} MR^2$$

PROBLEMS

1. Ceiling Fan: A simple model of a ceiling fan consists of a small cylindrical disk with five thin rods coming from the center. The disk has mass $m_d = 2.5\,kg$ and radius $R = 0.22$ m. The rods each have mass $m_r = 1.2$ kg and length $L = 0.7$ m. (a) What is the moment of inertia of each rod about the axis of rotation? (b) What is the moment of inertia of the disk about the axis of rotation? (c) What is the moment of inertia of the whole ceiling fan? (d) When the fan is turned on, it takes $t = 3.4$ s and a total of 15 revolutions to accelerate up to its full speed. What is the magnitude of the angular acceleration? (e) What is the final angular speed of the fan? (f) What is the final rotational energy of the fan? (g) The fan is turned to a lower setting where it ends with half of its rotational energy as before. The time it takes to slow to this new speed is also $t = 3.4$ s. What is the final angular speed of the fan? (h) What is the magnitude of the angular acceleration while the fan slows down?

FIGURE 14.8 Problem 1

2. Spinning Disk: A disk with mass $m = 11.8$ kg and radius $R = 0.36$ m begins at rest and accelerates uniformly for $t = 18.2$ s, to a final angular speed of $\omega = 30$ rad/s. The disk freely spins about its center. (a) What is the angular acceleration of the disk? (b) What is the angular displacement over the 18.2 s? (c) What is the moment of inertia of the disk? (d) What is the change in rotational energy of the disk? (e) What is the tangential component of the acceleration of a point on the rim of the disk when the disk has accelerated to half its final angular speed? (f) What is the radial component of the acceleration of a point on the rim of the disk when the disk has accelerated to half its final angular speed? (g) What is the final speed of a point on the disk half-way between the center of the disk and the rim? (h) What is the total distance a point on the rim of the disk travels during the 18.2 s?

UNIT

15

PARALLEL AXIS THEOREM AND TORQUE

15.1 Overview

In this unit we will continue our study of rotational motion. In particular, we will first prove a very useful theorem that relates moments of inertia about parallel axes. We will then move on to develop the equation that determines the dynamics for rotational motion. In so doing, we will introduce a new quantity, the torque, which plays the role for rotational motion that force does for translational motion. We will find, once again, that the rotational analog of mass will be the moment of inertia.

15.2 Parallel Axis Theorem

Earlier we showed that the total kinetic energy of a system of particles in any reference frame is equal to the kinetic energy of the center of mass of the system, defined to be one-half times the total mass times the square of the center-of-mass velocity, plus the kinetic energy of the motion of all of the parts relative to the center of mass.

$$K_{system,lab} = K_{REL} + K_{CM}$$

For a solid object the only possible motion relative to the center of mass is rotation. Therefore, the kinetic energy relative to the center of mass is just equal to one-half times the moment of inertia about a rotation axis, through the center of mass, times the square of the angular velocity about the center of mass.

$$K_{system,lab} = \frac{1}{2} I_{CM} \omega_{CM}^2 + \frac{1}{2} M v_{CM}^2$$

We can use this result to calculate the moment of inertia about a chosen axis if the moment of inertia about a parallel axis that passes through the center of mass is known.

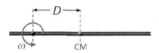

FIGURE 15.1 A thin rod rotates with angular velocity ω about an axis perpendicular to the page located a distance D from the center of mass of the rod.

For example, Figure 15.1 shows the thin rod from the last unit rotating about an axis perpendicular to the page at a distance D from the center of mass. We can describe this motion as the center of mass rotating about the axis with angular velocity, ω, plus the rod rotating about its center of mass with that same angular velocity, ω. Thus, for every rotation the center of mass makes about the axis, the rod makes one revolution about its center of mass.

We start from our previous result that the total kinetic energy about the chosen axis can be written in terms of the moment of inertia about that axis.

$$K_{Total} = \frac{1}{2} I_{Total} \omega^2$$

We know this kinetic energy must be equal to the sum of the kinetic energy of the center of mass and the kinetic energy relative to the center of mass:

$$K_{Total} = \frac{1}{2} M v_{CM}^2 + \frac{1}{2} I_{CM} \omega^2$$

The velocity of the center of mass in the lab frame is just equal to the product of D and ω, and the kinetic energy relative to the center of mass is just equal to one-half the product of the moment of inertia about this parallel axis and the square of the angular velocity.

$$K_{Total} = \frac{1}{2} M (D\omega)^2 + \frac{1}{2} I_{CM} \omega^2$$

We can cancel a factor of one-half ω^2 from each term in the kinetic energy equation to determine that the total moment of inertia about the chosen axis is just the moment of inertia about a parallel axis passing through the center of mass plus the moment of inertia of the center of mass, treated as a point particle, about the chosen axis.

$$I_{Total} = MD^2 + I_{CM}$$

This result is known as the **parallel axis theorem** and it will prove to be very useful in the next few units. You can verify that this parallel axis theorem predicts the relationship we obtained last time for the two moments of inertia for a thin rod.

15.3   Example: Moment of Inertia of a Dumbbell

We can now use this parallel axis theorem to calculate the moment of inertia of a dumbbell made up of two solid spheres connected by a solid rod about an axis that is perpendicular to the rod and passes through its center as shown in Figure 15.2.

We'll start by using our result from the last unit that the moment of inertia of the dumbbell about the given axis is equal to the sum of the moments of inertia of its components, the rod and the two spheres, about that same axis. To find the moments of inertia of the spheres about the axis through the center of the rod, we will apply the parallel axis theorem we developed in the last section.

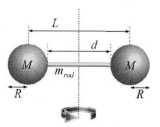

FIGURE 15.2 A dumbbell composed of two spherical masses of radius R and mass M separated by a distance d are connected by a rod with mass m_{rod}.

Namely, we know that the moment of inertia of a solid sphere about an axis passing through its center is equal to two-fifths the product of its mass and the square of its radius.

$$I_{sphere,CM} = \frac{2}{5} MR^2$$

Applying the parallel axis theorem, we see that the moment of inertia of each sphere about the given axis is just given by

$$I_{sphere,axis} = \frac{2}{5} MR^2 + M\left(\frac{L}{2}\right)^2$$

To find the total moment of inertia of the dumbbell, we just add the contributions from the two spheres to the moment of inertia of the thin rod about its center to get the result:

$$I_{dumbbell} = \frac{1}{12} m_{rod} d^2 + 2\left[\frac{2}{5} MR^2 + M\left(\frac{L}{2}\right)^2\right]$$

15.4   Torque and Angular Acceleration

So far in our study of rotations, we have made explicit connections to the kinematics of one-dimensional linear motion and the concept of mass. We will now use Newton's second law to develop the equation that describes the dynamics of rotations.

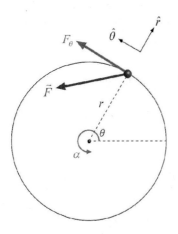

Figure 15.3 shows a point mass, constrained to move in a circle being acted upon by an arbitrary force $\vec{F}$. We choose to use polar coordinates (r and θ) to describe this motion since r will be constant. Since the mass is moving in a circle, we know there must be a component of the net force acting on the mass that is radial to supply the radial (centripetal) acceleration (v^2/r). In addition, there may be a component of the net force that is tangential which will result in a non-zero tangential acceleration. This tangential acceleration is simply related to the angular acceleration of the rotation.

$$F_\theta = ma_\theta = mr\alpha$$

FIGURE 15.3 A point mass constrained to move on a circle of radius r is acted upon by a force $\vec{F}$. The tangential component of this force, F_θ, gives rise to an angular acceleration α.

We know that the rotational analog of mass is the moment of inertia, which in this case is just equal to the product of the mass of the particle and the square of its distance from the axis of rotation.

$$I = mr^2$$

If we multiply both sides of our force equation by this distance r, we find that the product of r and the tangential component of the force is equal to the product of the moment of inertia of the object about the rotation axis and the angular acceleration.

$$rF_\theta = I\alpha$$

The right-hand side now looks like a rotational version of Newton's second law: namely, the product of the rotational mass (the moment of inertia) and the rotational acceleration. It would be natural to identify the left-hand side of this equation as the rotational force. Indeed, we will identify the product of the tangential force and the distance between the application point of this force and the rotation axis as the **torque**, the quantity that plays the role of force in rotational dynamics.

$$\tau \equiv rF_\theta$$

In the next sections we will formally define torque as a vector quantity and will arrive at a general vector equation that is the rotational version of Newton's second law.

15.5   Example: Closing a Door

We've just seen that a torque, the product of a tangential force and the distance between the application point of this force and the rotation axis, produces an angular acceleration. As a concrete example, Figure 15.4 shows the overhead view of an open door. To close this door, you need to push on it. Here, we see a force $\vec{F}$ being applied a distance r from the hinge. The door is heavy and barely moves as you push on it. What would you change about the way you are pushing in order to close it quicker? Your intuition tells you that you would either push harder, or you would push on the door at a point further from the hinge. You would certainly not try pushing closer to the hinge, and you would also not change the direction you were pushing to be more parallel to the door. In other words, the biggest effect you can have on the door is to push on it as hard as possible, in a direction perpendicular to the door, at a point far from the hinge.

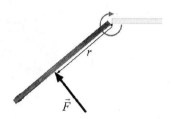

FIGURE 15.4 An overhead view of a closing door. The force $\vec{F}$ applies a torque about the hinge which causes an angular acceleration.

The equations we derived in the previous section tell us exactly the same thing! We expect the angular acceleration to be biggest when the torque is biggest–in other words the door will close fastest when the torque is large. The torque is largest when the distance r between the axis and the perpendicular force is biggest and when the perpendicular force itself is biggest. If we push parallel to the door, there is no perpendicular force component and we don't expect the door to start moving at all!

In the next section we will generalize this description of rotations by defining torque, angular velocity, and angular acceleration as vector quantities and by defining the cross product of two vectors.

15.6   Torque and the Cross Product

Figure 15.5 shows a spinning top. So far, we have described this motion in terms of one-dimensional variables, the angular displacement, the angular velocity, and the angular acceleration all defined relative to the axis of rotation. We will need to generalize this description to three dimensions when we allow the direction of the rotation axis itself to change.

If a rotating object is viewed in a reference frame in which the rotation axis is perpendicular to the page, as shown in Figure 15.6, it is conventional to define a counterclockwise rotation as positive and a clockwise rotation as negative. We adopt this convention in order

FIGURE 15.5 A top spins on its axis.

to match the usual measurement of the angle theta relative to the *x*-axis in a right-handed Cartesian coordinate system.

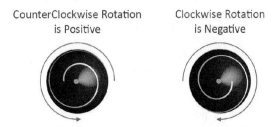

CounterClockwise Rotation is Positive   Clockwise Rotation is Negative

FIGURE 15.6 A spinning top viewed from above. A counterclockwise rotation is defined to be positive.

There is a simple rule involving the right hand that can be used to define the directions of the angular velocity vector. Namely, if you curl the fingers of your right hand in the direction of rotation of the object, your thumb will point in the direction of the angular velocity vector of the object (see Figure 15.7). In other words, if some object is spinning in the counterclockwise direction in the *x-y* plane, curling the fingers of your right hand in this direction results in your thumb pointing in the +*z* direction, which we define to be the direction of the angular velocity vector. Similarly, if the object is rotating in the clockwise direction, the same exercise results in your thumb pointing in the –*z* direction.

FIGURE 15.7 The *right-hand rule* for determining the direction of the angular velocity $\vec{\omega}$. Curl your fingers in the direction of rotation and your thumb will point along the direction of $\vec{\omega}$.

We use the same right-hand rule to define the direction of the angular acceleration. For example, if the magnitude of the angular velocity *increases* in time, then the angular acceleration vector has *the same* direction as that of the angular velocity. If the magnitude of the angular velocity *decreases* in time, then the angular acceleration vector has the *opposite* direction as that of the angular velocity.

One nice feature of using the right-hand rule to define the direction of vectors in systems involving rotation is that these vectors then do not rely on the choice of a coordinate system at all; the rotation vector becomes a property of the object itself.

Now that we have defined the direction of the angular velocity and angular acceleration vectors, we will now define the torque as a vector quantity and obtain the vector equation that determines rotational dynamics.

We have shown that the magnitude of the torque is given by the product of the length of the line connecting the axis to the point where the force acts, *r*, and the component of the force $\vec{F}$ that is perpendicular to this line. If we define θ as the angle between this line and the force as shown in Figure 15.8, then the perpendicular component of the force is just equal to the product of F and $\sin\theta$.

If we consider the line connecting the rotation axis to the force as a vector ($\vec{r}$), then the magnitude of the torque is given by the magnitude of the vector $\vec{F}$ times the magnitude of the vector $\vec{r}$ times the sine of the angle between them.

$$\tau = Fr\sin\theta$$

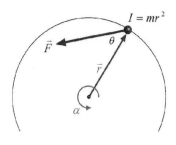

We write it in this way because there is a mathematical construct called the **cross product** that has exactly this same property. Namely, given two vectors $\vec{A}$ and $\vec{B}$, the magnitude of $\vec{A}\times\vec{B}$ is given by this same form:

FIGURE 15.8 The magnitude of the torque exerted by force $\vec{F}$ is equal to $rF\sin\theta$.

$$\left|\vec{A}\times\vec{B}\right| = AB\sin\theta$$

The direction of $\vec{A}\times\vec{B}$ is perpendicular to both $\vec{A}$ and $\vec{B}$ and is given by the right-hand rule shown in Figure 15.9. If you point the fingers of your right hand in the direction of $\vec{A}$ and curl them toward the direction of $\vec{B}$, then your thumb points in the direction of $\vec{A}\times\vec{B}$.

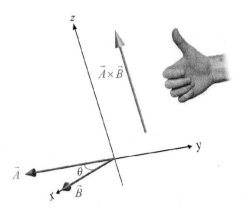

FIGURE 15.9 The direction of the cross product of two vectors $\vec{A}$ and $\vec{B}$ is given by a right-hand rule: point the fingers of your right hand in the direction of $\vec{A}$ and curl them toward the direction of $\vec{B}$, then your thumb points in the direction of $\vec{A}\times\vec{B}$.

Using this definition of the cross product, and what we already know about the right-hand rule, we can write a vector equation that neatly summarizes everything we know so far.

$$\vec{\tau}_{Net} = I\vec{\alpha}$$

On the right-hand side we have the angular acceleration vector, whose direction is found by applying the right-hand rule. On the left-hand side we have the torque vector, whose direction is also found using the right-hand rule.

In the example shown, we see that the directions of the torque and the angular acceleration are the same; they are both parallel to the rotation axis pointing out of the page. This equation is totally general: the net torque about an axis is equal to the product of the moment of inertia about that axis and the angular acceleration. This equation is just Newton's second law applied to a system of particles in rotation!

MAIN POINTS

Parallel Axis Theorem

The moment of inertia about a chosen axis is equal to the moment of inertia about a parallel axis passing through the center of mass plus the moment of inertia of the center of mass, treated as a point particle, about the chosen axis.

$$I_{Total} = I_{CM} + MD^2$$

Torque

The concept of torque plays the role for rotational motion that force does for translational motion.

$$\vec{\tau} \equiv \vec{r} \times \vec{F}$$

$$\tau = rF \sin \theta$$

The direction of the torque vector is determined by a right-hand rule: curl fingers from $\vec{r}$ into $\vec{F}$ and then thumb points in direction of torque.

Dynamics Equation for Rotational Motion

The equation that determines the dynamics of rotational motion is derived from Newton's second law.

$$\vec{\tau} = I\vec{\alpha}$$

The net torque about an axis is equal to the product of the moment of inertia about that axis and the angular acceleration.

PROBLEMS

1. Moment of Inertia: An object is formed by attaching a
uniform, thin rod with a mass of $m_r = 6.57$ kg and length
$L = 5.76$ m to a uniform sphere with mass $m_s = 32.85$ kg
and radius $R = 1.44$ m. Note $m_s = 5m_r$ and $L = 4R$. (a)
What is the moment of inertia of the object about an axis at
the left end of the rod? (b) If the object is fixed at the left
end of the rod, what is the angular acceleration if a force
$F = 439$ N is exerted perpendicular to the rod at the center
of the rod? (c) What is the moment of inertia of the object
about an axis at the center of mass of the object? (*Note*: The center of mass can be
calculated to be located at a point halfway between the center of the sphere and the left
edge of the sphere.) (d) If the object is fixed at the center of mass, what is the angular
acceleration if a force $F = 439$ N is exerted parallel to the rod at the end of rod? (e) What
is the moment of inertia of the object about an axis at the right edge of the sphere? (f)
Compare the three moments of inertia calculated above: $I_{CM} < I_{left} < I_{right}$,
$I_{CM} < I_{right} < I_{left}$, $I_{right} < I_{CM} < I_{left}$, $I_{CM} < I_{left} = I_{right}$, or $I_{right} = I_{CM} < I_{left}$.

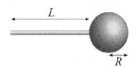

L

R

FIGURE 15.10 Problem 1

2. Torque on Disk: A uniform disk with
mass $m = 8.58$ kg and radius $R = 1.31$ m
lies in the x-y plane and is centered at the
origin. Three forces act in the $+y$
direction on the disk: 1) a force of 300 N
at the edge of the disk on the +x-axis, 2)
a force of 300 N at the edge of the disk
on the −y-axis, and 3) a force of 300 N
acts at the edge of the disk at an angle
$\theta = 37°$ above the −x-axis. (a) What is
the magnitude of the torque on the disk
due to F_1? (b) What is the magnitude of
the torque on the disk due to F_2? (c)
What is the magnitude of the torque on
the disk due to F_3? (d) What is the x
component of the net torque on the disk?
(e) What is the y component of the net
torque on the disk? (f) What is the z
component of the net torque on the disk?
(g) What is the magnitude of the angular
acceleration of the disk? (h) If the disk
starts from rest, what is the rotational
energy of the disk after the forces have been applied for $t = 1.7$ s?

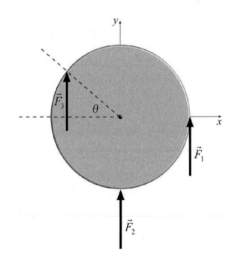

FIGURE 15.11 Problem 2

16

ROTATIONAL DYNAMICS

16.1   Overview

In this unit we will address examples that combine both translational and rotational motion. We will find that we will need both Newton's second law and the rotational dynamics equation we developed in the last unit to determine completely the motions. We will also develop the equation that is the rotational analog of the center-of-mass equation. Namely, we will find that the change in the rotational kinetic energy is determined by the integral of the torque over the angular displacement. We will close by examining in detail the motion of a ball rolling without slipping down a ramp.

16.2   Example: Disk and String

In the last unit we developed the vector equation that determines rotational dynamics, that the net torque on a system of particles about a given axis is equal to the product of the moment of inertia of the system about that axis and the angular acceleration.

$$\vec{\tau}_{Net} = I\vec{\alpha}$$

We will now apply this equation to a number of examples. We will start with the solid cylinder, mounted on a small frictionless shaft through its symmetry axis, as shown in Figure 16.1. It has a massless string wrapped around its outer surface. The string is pulled with a force $\vec{F}$ causing the cylinder to turn. Our task is to determine the resulting angular acceleration of the disk.

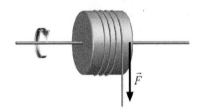

FIGURE 16.1 A force $\vec{F}$ is applied to a string wrapped around a solid cylinder mounted on a frictionless shaft producing an angular acceleration of the cylinder about its axis.

We will start by defining the system to be the disk and calculating the torque exerted on this system about the rotation axis. The torque is produced by the applied force $\vec{F}$ which always acts at a distance R from the axis. Furthermore, the direction of the force is always perpendicular to $\vec{R}$, the vector from the axis to the point of application of the force. Therefore, the torque vector ($\vec{R} \times \vec{F}$) has a magnitude equal to the product of R and F and a direction, obtained from the right-hand rule, that points along the axis to the right in the figure.

$$\vec{R} \times \vec{F} = I \vec{\alpha}$$

The direction of the angular acceleration must be the same as that of the torque. Consequently, since the disk was initially at rest, the disk rotates in the direction shown and its speed increases with time. Since we know the moment of inertia of a solid disk about its axis of symmetry ($I_{disk} = \frac{1}{2} MR^2$), we can solve for the magnitude of the angular acceleration.

$$\alpha = \frac{2F}{MR}$$

16.3   Combining Translational and Rotational Motion

Figure 16.2 shows the disk from the last section with a weight added to the end of the string. When we release the weight, the weight falls, pulling the string and causing the disk to rotate. In this example, we must deal with both the translational motion of the weight and the rotational motion of the disk. We want to calculate the resulting linear and angular accelerations.

How do we go about starting the calculation? To determine the motion of the weight, we will start by writing down Newton's second law. There are two forces acting on the weight: the tension force exerted by the string pointing up and the gravitational force exerted by the Earth pointing down. We will choose the positive y-axis to point down here which will result in a positive linear acceleration.

$$mg - T = ma$$

For the rotation of the disk, we have the same equation as before, with the applied force $\vec{F}$ replaced by the tension force $\vec{T}$.

$$RT = I\alpha$$

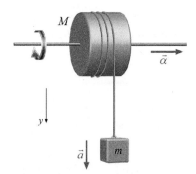

We now have two equations and three unknowns: the tension, and the linear and angular accelerations. We need another equation in order to solve the problem. The key here is to realize that since the string does not slip, the length of string that unwinds is equal to the arc length through which the disk turns! Therefore, we can use our result from the last unit that relates the linear acceleration of a point on the rim to the angular acceleration of the disk.

FIGURE 16.2 A mass m is attached to a string which is wrapped around a solid cylinder. As the mass falls, the string unwinds, producing an angular acceleration of the cylinder about its axis.

$$a = R\alpha$$

We now have three equations and three unknowns. All that is left to do is simply to solve these equations. We can first replace the angular acceleration in the rotational equation by the ratio of the linear acceleration to the radius of the disk to obtain

$$RT = I\frac{a}{R} \qquad \Rightarrow \qquad T = I\frac{a}{R^2}$$

We can now add this equation to Newton's second law equation for the weight in order to eliminate the tension.

$$mg = a\left(m + \frac{I}{R^2}\right)$$

We can then eliminate the moment of inertia by substituting in its value in terms of the mass and radius of the disk ($I_{disk} = \frac{1}{2}MR^2$) to obtain our result for the acceleration of the weight:

$$mg = a\left(m + \frac{1}{2}M\right) \qquad \Rightarrow \qquad a = g\left(\frac{m}{m + \frac{1}{2}M}\right)$$

We see that the acceleration of the weight is less than g by a factor determined by the masses of the weight and the disk. We can now use this value for the linear acceleration to determine the tension in the string:

$$T = m(g - a) = mg\left(\frac{M}{M + 2m}\right)$$

Here we see that the tension is less than the weight by another factor determined by the masses of the weight and the disk.

16.4   Work and Energy in Rotations

We now want to look at the rotational dynamics equation in the context of energy. Recall that by integrating Newton's second law for a system of particles, we obtained the center-of-mass equation, namely that the total macroscopic work done on the system is equal to the change in the center-of-mass kinetic energy, calculated as if the system were a point particle having the total mass of the system and moving with the velocity of the center of mass.

$$\int \vec{F}_{Net} \cdot d\vec{l}_{CM} = \Delta\left(\frac{1}{2}mv_{CM}^2\right)$$

We can obtain an exactly analogous equation for rotational motion relative to the center of mass. The derivation follows closely the previous derivation of the center-of-mass equation. Namely, if we replace the angular acceleration ($d\omega/dt$) in the rotational equation by the product of ω and $d\omega/d\theta$,

$$\alpha \equiv \frac{d\omega}{dt} = \frac{d\theta}{dt}\frac{d\omega}{d\theta} = \omega\frac{d\omega}{d\theta}$$

we obtain an equation that relates the net torque about an axis passing through the center of mass to the rate of change of the angular velocity to the angular displacement.

$$\tau_{Net} = I_{CM}\omega\frac{d\omega}{d\theta}$$

If we integrate this equation, we find the relationship we are looking for,

$$\int_{\theta_1}^{\theta_2} \tau_{Net}\, d\theta = \int_{\omega_1}^{\omega_2} I_{CM}\omega\, d\omega = \Delta\left(\frac{1}{2}I_{CM}\omega^2\right)$$

namely, that the integral of the torque over the angular displacement is equal to the change in the rotational kinetic energy. This relationship is completely general, and it will prove to be a powerful tool in solving rotational problems.

This result is actually more familiar than it might seem. For example, if we evaluate the integral of the torque over the angular displacement for the rotating disk in the last section, we find that it is just equal to the work done by the tension force!

$$\int_{\theta_1}^{\theta_2} \tau_{Net} \, d\theta = TR \int_{\theta_1}^{\theta_2} d\theta = TR \, \Delta\theta = TD$$

Namely, the torque is constant and equal to the product of the tension and the radius of the cylinder, while the change in angular displacement as the weight falls through a distance D is just equal to D/R. Consequently, we see that the integral of the net torque over the angular displacement is indeed equal to the product of the tension and the displacement of the weight which is just equal to the work done by the tension force!

16.5   Total Kinetic Energy of a Rolling Ball

We have previously shown that the total kinetic energy of a solid object is just equal to the kinetic energy of the center of mass of the object plus the kinetic energy due to the rotation of the object around an axis through the center of mass.

$$K_{Total} = \frac{1}{2} M_{Total} v_{CM}^2 + \frac{1}{2} I_{CM} \omega^2$$

The first term is called the translational kinetic energy of the object; the second term is called the rotational kinetic energy. For cases in which the object is rolling without slipping, we can simplify this expression since the angular velocity and the center-of-mass velocity are related in a very simple way.

Figure 16.3 shows the ball rolling through one revolution. As the ball rotates through an angular displacement θ, the center of mass moves through a distance equal to the arc length which is equal to the product of R and θ. Therefore, we see that the velocity of the center of mass is just equal to the product of the angular velocity of the ball and its radius!

We can now combine the kinetic energy of the center of mass with the kinetic energy of the rolling ball relative to the center of mass to obtain the total kinetic energy of the ball. Since the angular velocity of a ball that is rolling without slipping is simply related to its translational velocity, we can rewrite the total kinetic energy totally in terms of the ball's translational velocity.

$$K_{Total} = \frac{1}{2} M_{Total} v_{CM}^2 + \frac{1}{2} I_{CM} \left(\frac{v_{CM}}{R} \right)^2$$

Substituting in the moment of inertia for a solid sphere about its axis ($2M_{Total} R^2 / 5$) into this equation, we obtain our final result:

$$K_{Total} = \frac{7}{10} M_{Total} v_{CM}^2$$

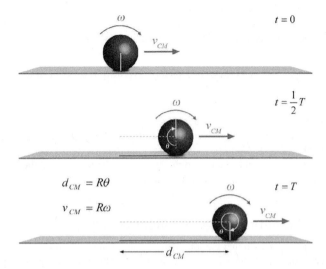

FIGURE 16.3 A ball rolls without slipping through a distance that corresponds to one complete revolution of the ball about its center. The center of the mass has traveled a distance $= v_{CM}t = v_{CM}(2\pi/\omega)$ which is also equal to $2\pi R$. Consequently, $v_{CM} = R\omega$.

Note that the total kinetic energy is now bigger than what it would be if the ball were sliding with the same speed. This is because we need to account for the additional kinetic energy due to rotation.

16.6   Ball Rolling Down a Ramp

We have just determined the total kinetic energy of a ball that rolls without slipping. We will now apply this result to the situation shown in Figure 16.4 where we see a solid sphere that is released from rest at the top of a ramp and then rolls without slipping to the bottom. Our task is to determine its speed when it reaches the bottom.

How do we go about solving this problem? We have certainly solved similar problems when the object was sliding, rather than rolling, down the ramp. In those cases, we applied the center-of-mass equation that says that the change in kinetic energy is equal to the macroscopic work done by all forces acting on the object:

$$\int \vec{F}_{Net} \cdot \vec{dl}_{CM} = \Delta\left(\frac{1}{2}mv_{CM}^2\right)$$

The center-of-mass equation still applies here: the macroscopic work done on the ball is equal to the product of the displacement of the center of mass and the difference between the component of the weight down the ramp and the frictional force, f.

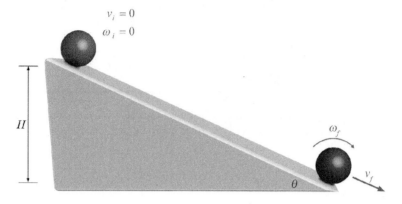

FIGURE 16.4 A ball is released from rest at the top of a ramp and rolls without slipping to the bottom. The speed of the ball at the bottom can be determined from energy considerations.

$$\int \vec{F}_{Net} \cdot d\vec{l}_{CM} = (mg \sin \theta - f)\Delta x_{CM} \qquad \rightarrow \qquad (mg \sin \theta - f)\Delta x_{CM} = \Delta \left(\frac{1}{2}mv_{CM}^2\right)$$

We cannot determine the final velocity of the center of mass from this equation, though, because we do not know the magnitude of the frictional force.

We can determine the magnitude of the frictional force, however, from a consideration of the rotational energy equation we recently derived, namely that the product of the net torque and the angular displacement is equal to the change in the rotational kinetic energy.

$$\int_{\theta_1}^{\theta_2} \tau_{Net} \, d\theta = \tau_{Net} \Delta \theta = \Delta \left(\frac{1}{2}I_{CM}\omega^2\right)$$

The net torque is just equal to the product of the frictional force and the radius of the ball.

$$\tau_{Net} = fR$$

We demonstrated in the last section that the product of the radius of the ball and the angular displacement is just equal to the displacement of the center of mass. Therefore, we can relate the change in the rotational kinetic energy of the ball to the product of the frictional force and the displacement of the center of mass.

$$f \Delta x_{CM} = \Delta \left(\frac{1}{2}I_{CM}\omega^2\right)$$

Combining this information with the center-of-mass equation, we obtain our final result: the change in the kinetic energy of the center of mass plus the change in the rotational kinetic energy relative to the center of mass is equal to the work done by the gravitational force.

$$\Delta\left(\frac{1}{2}mv_{CM}^2\right) + \Delta(\frac{1}{2}I_{CM}\omega^2) = mg\sin\theta\,\Delta x_{CM}$$

16.7   Acceleration of a Rolling Ball

In the last section we determined that the change in the kinetic energy of a ball rolling down a ramp was just equal to the work done by gravity. We would now like to determine the speed of the ball at any arbitrary time. In other words, we would like to calculate the acceleration of the ball.

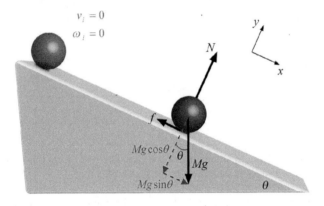

FIGURE 16.5 The free-body diagram for a ball rolling without slipping down a ramp.

We will start by drawing the free-body diagram as shown in Figure 16.5 and then writing Newton's second law that determines the motion of the center of mass.

$$Mg\sin\theta - f = Ma_x$$

We would like to solve this equation for the acceleration of the center of mass, but we don't know the magnitude of the frictional force! The only thing we know about the frictional force is that it is big enough to keep the ball from slipping.

The key to finding the magnitude of this force is to realize that it is the frictional force that supplies the torque that produces the angular acceleration of the ball. Therefore, the other equation we need is the rotational equation about the center of mass of the ball ($\vec{\tau}_{Net} = I_{CM}\vec{\alpha}_{CM}$). The magnitude of the torque is just equal to the product of the radius of the ball and the frictional force. The direction of the torque is into the page.

$$fR = I_{CM}\alpha_{CM}$$

Therefore, the angular acceleration of the ball is just equal to the magnitude of the torque divided by the moment of inertia about an axis passing through the center of mass of the ball.

$$\alpha_{CM} = \frac{fR}{I_{CM}}$$

Looking at the two equations we now have (Newton's second law for translations and rotations), we see we have three unknowns: the frictional force, and the linear and angular accelerations. We need to eliminate one more unknown and that we can do because we know that in rolling without slipping, the angular acceleration about the center of mass is just equal to the acceleration of the center of mass divided by the radius of the ball.

$$\alpha_{CM} = \frac{a_x}{R}$$

Making this substitution for the angular acceleration, we can now solve for both the magnitude of the frictional force and the acceleration of the center of mass.

$$\frac{a_x}{R} = \frac{fR}{I_{CM}} = \frac{f}{\frac{2}{5}MR} \qquad \Rightarrow \qquad f = \frac{2}{5}Ma_x$$

$$Mg\sin\theta - f = Ma_x$$

We find that the acceleration of the center of mass is smaller than that of an object sliding down a frictionless ramp inclined at the same angle (which was equal to $g\sin\theta$).

$$a_x = \frac{5}{7}g\sin\theta$$

16.8   Why Did that Last Derivation Work?

We just derived the acceleration of a ball rolling without slipping down a ramp by applying the rotational equation ($\vec{\tau}_{Net} = I\vec{\alpha}$) in which we evaluated the torque about an axis passing through the center of mass of the ball. Recall that we obtained the rotational equation from Newton's second law and that Newton's second law is valid in inertial reference frames. The reference frame of the ball is clearly *not* an inertial reference frame. What is going on here?

The surprising answer to this question is that we can *always* apply this rotational equation, even for an object that is accelerating, as long as we are considering only rotations about the center of mass of the object! This result is certainly *not* obvious and its proof requires the introduction of a concept, angular momentum, which we will discuss in a future unit. For the benefit of those of you who are curious, we will present a proof of this claim now.

We will start with the definition of the **angular momentum** $\vec{L}$ of a particle about some axis as the cross product of $\vec{r}$, the vector from the axis to the particle, with the momentum vector:

$$\vec{L} \equiv \vec{r} \times \vec{p}$$

Taking the derivative of $\vec{L}$ with respect to time, we get two terms.

$$\frac{d\vec{L}}{dt} = \left(\frac{d\vec{r}}{dt} \times \vec{p}\right) + \left(\vec{r} \times \frac{d\vec{p}}{dt}\right)$$

The first term is zero since the velocity vector ($d\vec{r}/dt$) and the momentum vector are parallel. We can use Newton's second law ($\vec{F}_{Net} = d\vec{p}/dt$) to write the second term as the net torque on the particle.

$$\frac{d\vec{L}}{dt} = \left(\frac{d\vec{r}}{dt} \times \vec{p}\right) + \left(\vec{r} \times \frac{d\vec{p}}{dt}\right) = 0 + \vec{r} \times \vec{F}_{Net} = \vec{\tau}_{Net}$$

We will use this equation, which determines the time dependence of the angular momentum vector of the particle, to show that the rotational equation ($\vec{\tau}_{Net} = I\vec{\alpha}$) holds in the frame of the ball that is rolling without slipping down the ramp.

The first step is to obtain an important result concerning the angular momentum of a system of particles, namely

$$\vec{L} = \vec{L}_{CM} + \vec{L}^{*}$$

Where $\vec{L}$ is the total angular momentum of a system of particles in a particular reference frame, $\vec{L}_{CM}$ is the angular momentum of the center of mass of the system in that frame, and $\vec{L}^{*}$ is the angular momentum of the system in the center-of-mass frame. The proof is straightforward but a little lengthy. We start by expressing the displacement vector is the specified frame in terms of the displacement vector of the center of mass of the system in that frame and the displacement vector in the center-of-mass frame.

$$\vec{r}_i = \vec{r}_{CM} + \vec{r}_i^{*}$$

We now write down the expression for the angular momentum of a system of particles:

$$\vec{L} = \sum \left[\left(\vec{r}_{CM} + \vec{r}_i^{*}\right) \times \vec{p}_i \right] = \left(\vec{r}_{CM} \times \sum \vec{p}_i\right) + \sum \left(\vec{r}_i^{*} \times \vec{p}_i\right)$$

The first term on the right-hand side of the above equation is equal to the angular momentum of the center of mass since the sum of all the individual momenta is just the total momentum of the system!

$$\vec{L}_{CM} = \vec{r}_{CM} \times \sum \vec{p}_i = \vec{r}_{CM} \times \vec{P}_{Total}$$

The second term on the right-hand side must be the angular momentum of the system in the center-of-mass frame. To demonstrate this claim, we need to first expand the individual momenta in the specified frame in terms of the individual momenta in the center-of-mass frame.

$$\vec{p}_i = m_i \frac{d\vec{r}_i}{dt} = m_i \left(\frac{d}{dt}\vec{r}_{CM} + \frac{d}{dt}\vec{r}_i^{\,*} \right) = m_i\vec{v}_{CM} + m_i \frac{d\vec{r}_i^{\,*}}{dt}$$

$$\sum \left(\vec{r}_i^{\,*} \times \vec{p}_i \right) = \sum \left(\vec{r}_i^{\,*} \times m_i\vec{v}_{CM} \right) + \sum \left(\vec{r}_i^{\,*} \times m_i\vec{v}_i^{\,*} \right)$$

The first term on the right-hand side of the last equation is zero since we can take the velocity of the center-of-mass vector outside the sum, leaving the mass-weighted sum of the displacements in the center of mass which must just be zero!

$$\sum \left(\vec{r}_i^{\,*} \times m_i\vec{v}_{CM} \right) = \left(\sum m_i\vec{r}_i^{\,*} \right) \times \vec{v}_{CM} = 0$$

We are left then with our result:

$$\sum \left(\vec{r}_i^{\,*} \times \vec{p}_i \right) = \sum \left(\vec{r}_i^{\,*} \times m_i\vec{v}_i^{\,*} \right) = \vec{L}^{\,*} \qquad \Rightarrow \qquad \vec{L} = \vec{L}_{CM} + \vec{L}^{\,*}$$

We are now finally ready to prove that we can *always* apply the rotational equation ($\vec{\tau}_{Net} = I\vec{\alpha}$), even for an object that is accelerating, as long as we are considering only rotations about the center of mass of the object!

Suppose we fix our *x-y* coordinate system to the ramp as shown in Figure 16.6. This system is clearly an inertial reference frame; therefore, our angular momentum equation holds in this frame. Considering the ball to be a system of particles, we can write down the equation of motion in terms of the time rate of change of the angular momentum of the system:

$$\sum \vec{\tau}_i = \sum \frac{d\vec{L}_i}{dt}$$

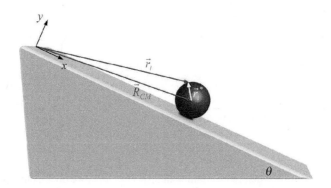

FIGURE 16.6 The displacement vector $\vec{r}_i$ in the inertial reference frame fixed to the ramp (*x-y*) is equal to the vector sum of the displacement vector of the center of mass $\vec{R}_{CM}$ plus the displacement vector relative to the center of mass $\vec{r}_i^{\,*}$.

We will now expand the torques in terms of the individual displacements:

$$\sum \vec{\tau}_i = \sum \left(\vec{r}_i \times \vec{F}_i \right) = \sum \left[\left(\vec{r}_{CM} + \vec{r}_i^* \right) \times \vec{F}_i \right]$$

Similarly, we can expand the angular momentum of the system:

$$\sum \frac{d\vec{L}_i}{dt} = \dot{\vec{L}}_{CM} + \dot{\vec{L}}^*$$

Putting this altogether, we obtain the equation:

$$\sum \left[\left(\vec{r}_{CM} + \vec{r}_i^* \right) \times \vec{F}_i \right] = \frac{d}{dt} \left(\vec{L}_{CM} + \vec{L}^* \right)$$

We now have the equation we need. We just have to expand the terms and simplify. Unfortunately, once again this process is a little lengthy. We start by expanding the time rate of change of the angular momentum of the center-of-mass term:

$$\frac{d}{dt} \vec{L}_{CM} = \frac{d}{dt} \left(\vec{r}_{CM} \times M_{Total} \vec{v}_{CM} \right) = \left(\vec{v}_{CM} \times M_{Total} \vec{v}_{CM} \right) + \left(\vec{r}_{CM} \times M_{Total} \vec{a}_{CM} \right)$$

The first term on the right-hand side of the last equation is zero since the cross product of any vector with itself is zero. The second term is equal to the net force on the system from Newton's second law. Therefore, we have obtained the expression for the time rate of change of the angular momentum of the center of mass:

$$\frac{d}{dt} \vec{L}_{CM} = \left(\vec{r}_{CM} \times \sum \vec{F}_i \right)$$

If we now substitute this form back into our master equation (three above), we get cancellations that leave us with

$$\sum \left(\vec{r}_i^* \times \vec{F}_i \right) = \frac{d}{dt} \vec{L}^*$$

We have now obtained our result: the sum of the torques about an axis through the center of mass is equal to the time rate of change of angular momentum of the system in the center-of-mass frame. For a rigid, symmetric, solid object (such as our rolling ball),

$$\vec{L}^* = I_{CM} \vec{\omega}_{CM}$$

Differentiating this angular momentum with respect to time, we obtain our result:

$$\frac{d}{dt} \vec{L}^* = I_{CM} \frac{d}{dt} \vec{\omega}_{CM} = I_{CM} \vec{\alpha} \qquad \Rightarrow \qquad \vec{\tau}_{CM} = I_{CM} \vec{\alpha}$$

It's been a long haul, but we have finally proved the important result that the sum of the torques about the center of mass of a system of particles is always equal to the product of the moment of inertia about the center of mass and the angular acceleration about the center of mass, even if the system itself is accelerating!

MAIN POINTS

Center of Mass Equation for Rotational Motion

Integrating the rotational dynamics equation, we determined that the change in the rotational kinetic energy is equal to the integral of the torque over the angular displacement.

$$\int_{\theta_i}^{\theta_f} \tau_{Net} \, d\theta = \Delta\left(\frac{1}{2} I_{CM} \omega^2\right)$$

Translation + Rotation: Rolling Without Slipping

The total kinetic energy of an object is equal to the sum of the translational kinetic energy of the center of mass and the rotational motion about the center of mass.

$$K_{Total} = \frac{1}{2} M v_{CM}^2 + \frac{1}{2} I_{CM} \omega^2$$

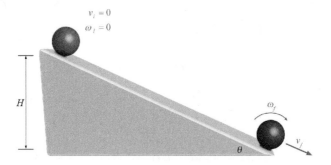

Applying the center-of-mass equation and its rotational analog, we determined that the change in the total kinetic energy of a solid ball rolling without slipping down the ramp is equal to the work done by gravity.

$$\Delta\left(\frac{1}{2} M v_{CM}^2\right) + \Delta\left(\frac{1}{2} I_{CM} \omega^2\right) = W_{gravity}$$

Applying Newton's second law and the rotational dynamics equation, we determined the acceleration of a solid ball down the ramp.

$$a_{CM} = \frac{5}{7} g \sin \theta$$

PROBLEMS

1. Bowling Ball: A spherical bowling ball with mass $m = 3.2$ kg and radius $R = 0.103$ m is thrown down the lane with an initial speed of $v = 8.5$ m/s. The coefficient of kinetic friction between the sliding ball and the ground is $\mu = 0.35$. Once the ball begins to roll without slipping, it moves with a constant velocity down the lane. (a) What is the magnitude of the angular acceleration of the bowling ball as it slides down the lane? (b) What is magnitude of the linear acceleration of the bowling ball as it slides down the lane? (c) How long does it take the bowling ball to begin rolling without slipping? (d) How far does the bowling ball slide before it begins to roll without slipping? (e) What is the magnitude of the final velocity? (f) After the bowling ball begins to roll without slipping, compare the rotational and translational kinetic energy of the bowling ball: $K_{rot} < K_{tran}$, $K_{rot} = K_{tran}$, or $K_{rot} > K_{tran}$.

2. Three Masses: A hoop with mass $m_h = 2.5$ kg and radius $R_h = 0.16$ m hangs from a string that goes over a solid disk pulley with mass $m_d = 2.1$ kg and radius $R_d = 0.09$ m. The other end of the string is attached to a massless axel through the center of a sphere on a flat horizontal surface that rolls without slipping and has mass $m_s = 4.1$ kg and radius $R_s = 0.18$ m. The system is released from rest.

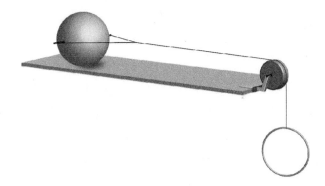

FIGURE 16.7 Problem 2

(a) What is magnitude of the linear acceleration of the hoop? (b) What is the magnitude of the linear acceleration of the sphere? (c) What is the magnitude of the angular acceleration of the disk pulley? (d) What is the magnitude of the angular acceleration of the sphere? (e) What is the tension in the string between the sphere and disk pulley? (f) What is the tension in the string between the hoop and disk pulley? (g) The hoop falls a distance $d = 1.6$ m after being released from rest. How much time does the hoop take to fall 1.6 m? (h) What is the magnitude of the velocity of the hoop after it has dropped 1.6 m? (i) What is the magnitude of the final angular speed of the sphere after the green hoop has fallen the 1.6 m?

3. Disk and String (INTERACTIVE EXAMPLE): A puck rests on a horizontal frictionless plane. A string is wound around the puck and pulled with constant force. What fraction of the disk's total kinetic energy is due to the rotation, K_{rot} / K_{total} ?

4. Sphere Incline (INTERACTIVE EXAMPLE): A solid sphere of uniform density starts from rest and rolls without slipping down an inclined plane with angle $\theta = 30°$. The sphere has mass $M = 8$ kg and radius $R = 0.19$ m. The coefficient of static friction between the sphere and the plane is $\mu = 0.64$. What is the magnitude of the frictional force on the sphere?

17

ROTATIONAL STATICS: PART I

17.1 Overview

In this unit we will focus on statics, the study of forces on objects that do not move. We will start by looking at the torque produced by the weight of a solid object and will find that it can be calculated by simply assuming that the entire mass of the object is located at its center of mass. We will then move on to determine the general conditions that must be satisfied for a system to be in static equilibrium. First, the sum of all the forces acting on the system must be zero. Second, the sum of the torques produced by the forces acting on the system must also be equal to zero. We will close by applying these conditions to a few specific examples.

17.2   Torque Due to Gravity

Up to this point we have considered the torque due to forces whose point of application was clear. The tension in a string wrapped around a cylinder provides a force acting at the edge of the cylinder, as does the force of friction acting on a ball rolling down a ramp. We now want to consider torques due to continuous mass distributions. We will show that the torque produced by the weight of a solid object can be calculated by simply assuming that the entire mass of the object is located at its center of mass.

Figure 17.1 shows a thin beam attached at one end to a wall by a hinge. The hinge can rotate freely, so when the beam is released it will start to swing downward. Clearly, the gravitational force is producing the torque about the rotation axis defined by the hinge.

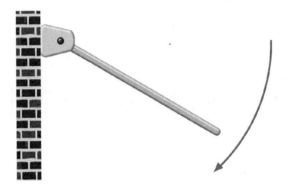

FIGURE 17.1 A thin beam is attached to a wall by a hinge. When released, the beam rotates about the hinge.

To calculate this torque, we need to integrate the contributions due to all of the mass elements dm along the length of the beam around this axis as shown. Each mass element will produce a torque into the page so that all we need to do is to set up an integral to determine the magnitude of the torque. This integral involves a cross product which can be expanded in terms of the angle θ between the beam and the vertical as shown in Figure 17.2.

$$\tau = \int d\tau = \int r(dm\,g)\sin\theta$$

After taking the constant terms outside the integral, we are left with the integral of $r\,dm$ over the length of the beam. This integral is exactly the integral that is needed to determine the position of the center of mass.

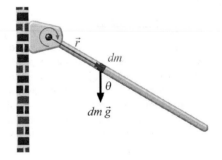

FIGURE 17.2 The torque generated by a mass element dm points into the page with magnitude $d\tau = r(dm\,g)\sin\theta$.

$$R_{CM} \equiv \frac{1}{M} \int r \, dm$$

Therefore, we can replace the integral in our expression for the magnitude of the torque by the distance from the hinge to the center of mass.

$$\tau = R_{CM} Mg \sin \theta$$

The resulting expression for the magnitude of the torque is, as promised, just equal to what we would obtain for a point particle located at the center of mass of the beam and whose mass is equal to the mass of the beam.

$$\vec{\tau}_{gravity} = \vec{R}_{CM} \times M\vec{g}$$

This result is completely general and is true for any shape. We chose a thin beam for this example to simplify the calculation.

17.3   Torque and Center-of-Mass Displacement from the Pivot

In the last section we found that the torque on a solid object about some rotation axis due to gravity can be found by assuming that the weight of the object acts at the center of mass of the object. We will now examine this result in more detail to obtain an alternate expression for the torque.

From the definition of the cross product we know that the magnitude of the torque can be expressed in terms of the magnitudes of $\vec{R}_{CM}$, the displacement vector of the center of mass relative to the rotation axis, the weight, and the angle between these two vectors.

$$\left| \vec{\tau}_{gravity} \right| = \left| \vec{R}_{CM} \times M\vec{g} \right| = R_{CM} Mg \sin \theta$$

Since the weight always points down, the product of R_{CM} and the sine of this angle will always be equal to the horizontal displacement of the center of mass from the rotation axis. In other words, the magnitude of the torque on some object due to gravity about any rotation axis is just the product of the weight of the object and the horizontal distance that the center of mass is displaced from a vertical line through the axis as shown in Figure 17.3.

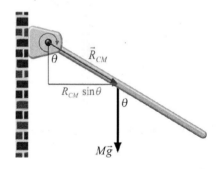

FIGURE 17.3 The magnitude of the torque due to gravity of the beam is equal to the product of the weight of the beam and the horizontal distance of the center of mass of the beam to the rotation axis.

$$\left| \vec{\tau}_{gravity} \right| = Mg \left(R_{CM} \sin \theta \right)$$

This result does make sense. For example, the torque due to gravity will be zero if the center of mass is directly below the axis; the object will just hang there without moving. This result also implies that the torque is biggest when the object is rotated so that the center of mass is at the same vertical height as the axis, making the sine of the angle between $\vec{R}_{CM}$ and the weight its maximum value of one.

It is important to note that the torque will *always* have the form we see here; it can always be expressed as a product of a force and the perpendicular distance of a line through this force and the rotation axis. This perpendicular distance, $r_{\perp}$, is called the **lever arm**.

$$\tau = r_{\perp} F \equiv (r \sin \theta) F \qquad \text{(where } \theta \text{ is the angle between } \vec{r} \text{ and } \vec{F} \text{)}$$

In the case of gravity we see that the lever arm is just the horizontal displacement of the center of mass from the rotation axis. In general the lever arm does not have to be horizontal, but it will always be perpendicular to the force. We will consider an example in the next section that illustrates this calculation.

17.4   Statics Example: Beam and Wire

Figure 17.4 shows a horizontal beam having mass M and length L attached to a vertical wall by a hinge. The far end of the beam is supported by a massless wire that makes an angle θ with the beam. Our task is to determine the tension in the wire.

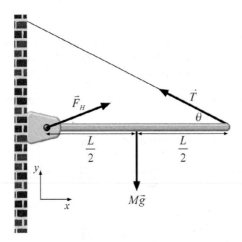

FIGURE 17.4 A beam of mass M and length L is supported by a massless wire that makes an angle θ with the beam. The task is to calculate the tension in the wire.

We begin by drawing and labeling all forces acting on the beam: The weight acts at the center of mass and the tension due to the wire acts at the end of the beam in the direction of the wire. There is also a third force due to the hinge, which supports the beam at the wall. We will call this force $\vec{F}_H$ and allow it to have both x and y components.

Since the system is static, both the acceleration of the center of mass and the angular acceleration about any rotation axis must be zero. Since the acceleration of the center of mass of the beam is zero, the total force on the beam must be zero in both the x and y directions. Therefore, we can apply Newton's second law to obtain two equations:

$$F_{H,x} - T\cos\theta = 0$$
$$F_{H,y} + T\sin\theta - Mg = 0$$

These two equations alone do not allow us to solve for the tension in the string since there are three unknowns: the tension and both of the components of the force due to the hinge. The good news is that we can get another equation by realizing that the angular acceleration of the beam about any axis is zero, which means that the torque about any axis must also be zero.

We now want to choose an axis of rotation for the calculation of the torque that simplifies our calculation. Since our task is just to find the tension in the wire, we can pick a rotation axis perpendicular to the page that passes through the hinge. We see immediately that the lever arm of the force due to the hinge is zero about this axis, so the torque due to this force about this axis is also zero.

What about the torques due to the other two forces? The magnitude of the torque due to the weight of the beam is just equal to the product of the weight and its lever arm. Using the right-hand rule, we see that the direction of this torque is into the page. The magnitude of the torque due to the wire is equal to the product of the tension in the wire and its lever arm. Using the right-hand rule, we see that the direction of this torque is out of the page. Therefore, the two torques are in opposite directions, and we assign them opposite signs. Setting the sum of these torques equal to zero, we obtain

$$TL\sin\theta - Mg\frac{L}{2} = 0$$

We can solve this equation to find a simple relationship between the tension in the wire, the weight of the beam, and the angle of the wire.

$$T = \frac{Mg}{2\sin\theta}$$

We can now determine the force the hinge exerts on the beam by solving our two original equations we found using Newton's second law after substituting in our newly found value for the tension in the wire,

$$F_{H,x} = \frac{Mg}{2\tan\theta} \qquad\qquad F_{H,y} = \frac{Mg}{2}$$

17.5 Example: Change Angle of Beam

We will now change the geometry of our beam and wire example as shown in Figure 17.5 and determine the force at the hinge as well as the tension in the wire. Once again we start by writing down Newton's second law for both the x and y components.

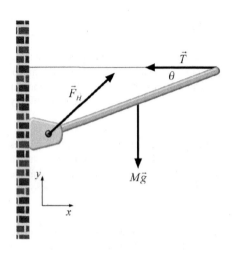

$$F_{H,x} - T = 0$$
$$F_{H,y} - Mg = 0$$

We can immediately solve the y equation for $F_{H,y}$.

$$F_{H,y} = Mg$$

The x equation, though, contains two unknowns, the tension and the x component of the force exerted by the hinge. To solve for these unknowns, we need to write down the torque equation. We again pick a rotation axis through the hinge, making the torque due to the force of the hinge zero. The magnitude of the torque due to the weight of the beam is once again equal to the product of the weight and its lever arm. Using the right-hand rule we see that the direction of this torque is into the page. The magnitude of the torque due to the wire is equal to the product of the tension and its lever arm. Using the right-hand rule we see that the

FIGURE 17.5 A beam of mass M and length L is supported by a horizontal massless wire. The beam makes an angle θ with the wire. The task is to calculate the tension in the wire and the force at the hinge.

direction of this torque is out of the page. These two torques are once again in opposite directions so we assign them opposite signs.

$$TL\sin\theta - Mg\frac{L}{2}\cos\theta = 0$$

We can solve this equation to find a simple relationship between the tension in the wire, the weight of the beam, and the angle of the wire.

$$T = \frac{Mg}{2\tan\theta}$$

We can now return to Newton's second law equations to determine the horizontal component of the force exerted by the hinge on the beam.

$$F_{H,x} = \frac{Mg}{2\tan\theta}$$

The procedure we have demonstrated in these last two examples will work well for any statics problem. Namely, simply start by drawing a picture that shows all forces and where they act. Using this picture, you can write down equations that require the translational and angular accelerations to be zero. You are then free to choose a rotation axis that simplifies the calculation of the torques.

17.6 Example: Wire Breaks

We will close this unit by considering what happens when the wire in the previous example breaks. What is the angular acceleration of the beam as it falls? Figure 17.6 shows the string breaking and the beam rotating about the hinge. This motion is a pure rotation, whose angular acceleration can be determined using the torque equation.

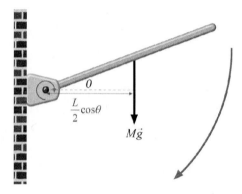

FIGURE 17.6 The wire in Figure 17.5 breaks and the beam rotates about the hinge. The task is to calculate the resulting angular acceleration.

Since the wire is no longer attached to the beam, the only torque around this rotation axis is that due to gravity. In the last section we determined both the magnitude and the direction of this torque.

$$\left|\vec{\tau}_{gravity}\right| = Mg\frac{L}{2}\cos\theta \qquad \text{(Direction is into page)}$$

We can therefore determine the magnitude of the angular acceleration by simply dividing this torque by the moment of inertia of the beam around the rotation axis. Previously, we determined that the moment of inertia of a beam about its end is equal to one third its mass times the square of its length ($I_{beam\ at\ end} = ML^2/3$).

$$\tau_{Net} = I\alpha \qquad\qquad \rightarrow \qquad\qquad \alpha = \frac{3g}{2L}\cos\theta$$

The first thing to note about our result is that the angular acceleration is not constant; it depends on the angle θ. It is biggest when the beam is horizontal and smallest when it is vertical. Of course, the angular displacement and the angular acceleration are not independent variables. If we write α in terms of the second derivative of θ with respect to time, we obtain the differential equation:

$$\frac{3g}{2L}\cos\theta = -\frac{d^2\theta}{dt^2}$$

The minus sign enters because a positive angular acceleration corresponds to an increasing angular velocity which corresponds to an angular acceleration into the page. This equation is difficult to solve analytically, but we can, and we will, solve it in the small angle approximation in a few units when we discuss simple harmonic motion.

MAIN POINTS

Torque Due to Weight of Object

The torque produced by the weight of a solid object can be calculated by simply assuming the entire mass of the object is located at its center of mass.

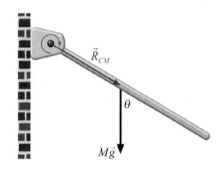

$$\vec{\tau}_{gravity} = \vec{R}_{CM} \times M\vec{g}$$

Lever Arm Calculation of Torque

The magnitude of any torque can be calculated as the product of the force and its lever arm, the perpendicular distance of a line through this force and the rotation axis.

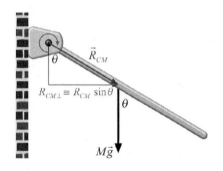

$$\tau = r_\perp F \equiv (r\sin\theta)F$$

Conditions for Static Equilibrium

The condition for static equilibrium is that the acceleration of the center of mass and the angular acceleration about any rotation axis must be zero.

Therefore, for a system to be in static equilibrium, the sum of the forces acting on the system must be zero *and* the sum of all torques produced by these forces must also be zero.

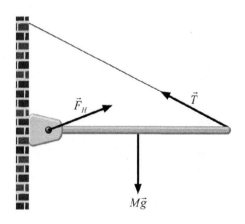

$$\sum \vec{F}_i = m\vec{a}_{CM} = 0$$

$$\sum \vec{\tau}_i = I\vec{\alpha} = 0 \quad \text{(about any axis)}$$

PROBLEMS

1. Hanging Beam: A beam is hinged to a wall to hold up a block. The beam has a mass of $m_{beam} = 6.9$ kg and the block has a mass of $m_{block} = 17.4$ kg. The length of the beam is $L = 2.46$ m. The block is attached at the very end of the beam, but the horizontal wire holding up the beam is attached two-thirds of the way to the end of the beam. The angle the wire makes with the beam is $\theta = 30°$. (a) What is the tension in the wire that holds the beam in place? (b) What is the net force the hinge exerts on the beam? (c) The maximum tension the horizontal wire can have without breaking is $T = 1,042$ N. What is the maximum mass a block can have and still be hung from the end of the beam? (d) Which of the following could be done in order to hold a heavier block?

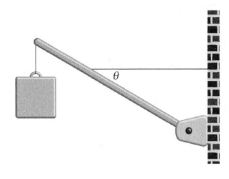

FIGURE 17.7 Problem 1

   (i)   Attach the wire to the end of the beam while still keeping it horizontal.
   (ii) Make the wire perpendicular to the beam while keeping the wire attached at the same location on the beam.
   (iii) Attach the block on the beam closer to the wall.
   (iv) Shorten the length of the wire attaching the block to the beam.

2. Gymnast: A gymnast with mass $m_{gym} = 50$ kg is on a balance beam that sits on (but is not attached to) two supports. The beam has a mass $m_{beam} = 124$ kg and length $L = 5$ m. Each support is one-third of the way from each end. Initially the gymnast stands at the left end of the beam. (a) What is the force the left support exerts on the beam? (b) What is the force the right support exerts on the beam? (c) How much extra mass could the gymnast hold before the beam begins to tip? (d) The gymnast (not holding any additional mass) walks directly above the right support. What is the force the left support exerts on the beam? (e) What is the force the right support exerts on the beam? (f) At what location does the gymnast need to stand to maximize the force on the right support: *at the center of the beam, at the right support,* or *at the right edge of the beam*?

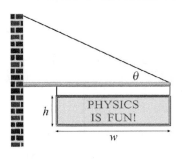

FIGURE 17.8 Problem 3

3. Sign (INTERACTIVE EXAMPLE): A sign has a mass of 1,050 kg, a height $h = 1$ m, and a width $w = 4$ m. It is held by a light rod of length 5 m that is perpendicular to a rough wall. A guy wire at $\theta = 23°$ to the horizontal holds the sign to the wall. Note that the distance from the left edge of the sign to the wall is 1 m. Suppose we rely upon friction between the wall and the rod to hold up the sign (there is no hinge attaching the rod to the wall). What is the smallest value of the coefficient of friction μ such that the sign will remain in place?

18

ROTATIONAL STATICS: PART II

18.1 Overview

In this unit we will conclude our study of statics. We will first work through a couple of examples in some detail. In particular, we will determine the maximum angle that a ladder can make with the wall before it starts to slip, and we will also determine the maximum acceleration a truck can have before a box in its bed tips over. We will then move on to explore the conditions for static equilibrium of an object considered from two perspectives: the torque produced by the weight of the object and the gravitational potential energy of the object. We will close with some remarks about viewing the stability of an object in terms of the position of its center of mass relative to its footprint.

18.2 Example: Ladder

In the last unit we introduced a procedure for solving problems where the object under study is not moving. Since these kinds of problems deal with objects in static equilibrium, this field of study is often referred to as *statics*. The approach for solving all statics problems is more or less the same. Start by drawing a free-body diagram of the object under study, and then use the fact that the total force on the object must be zero in any direction and the total torque on the object about any axis must also be zero to obtain the equations needed to solve for the unknowns.

We will now illustrate this procedure by solving a classic statics problem: that of a ladder of length L and mass M leaning against a wall, as shown in Figure 18.1. We assume there is no friction between the ladder and the wall, but there is friction between the ladder and the ground characterized by a known coefficient of static friction. Our job is to determine the maximum angle that the ladder can make with the wall before it starts to slip and fall down.

We start by drawing the free-body diagram of the ladder. The weight of the ladder acts at its center of mass. At the point of contact between the ladder and the ground there is both a normal force which points upward and a horizontal frictional force that points toward the wall. At the point of contact between the ladder and the wall there is only a horizontal normal force exerted by the wall on the ladder, since we have assumed no friction between the ladder and the wall.

Once we have drawn this free-body diagram, we then set the sum of all forces acting on the ladder equal to zero. This procedure creates two equations. In the vertical direction, we see that the magnitude of the normal force exerted by the ground must be equal to the weight of the ladder.

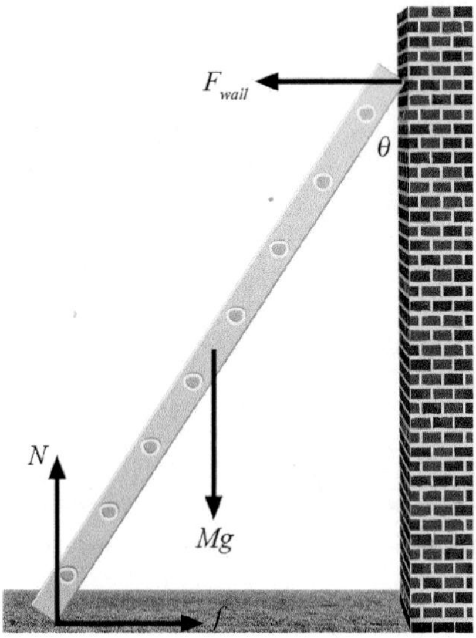

FIGURE 18.1 A ladder of mass M and length L rests against a wall. All forces acting on the ladder are shown, assuming that there is friction between the ladder and the ground but not between the ladder and the wall.

$$N - Mg = 0$$

In the horizontal direction, we see that the magnitude of the force exerted by the wall on the ladder must be equal to the frictional force exerted by the ground on the ladder.

$$f - F_{wall} = 0$$

The ladder will just begin to slip when the frictional force has its maximum value. Therefore, we can set the frictional force in the above equation equal to its maximum value.

$$f \rightarrow f_{max} = \mu_S N$$

We now know the magnitudes of all the forces.

$$N = Mg$$

$$F_{wall} = f = \mu_S Mg$$

To find the maximum angle that the ladder can make with the wall before it starts to slip, we need to choose a rotation axis and then set the sum of all torques about that axis equal to zero. We will choose an axis perpendicular to the page that passes through the point of contact between the ladder and the ground in order to minimize the number of terms in the torque equation. The weight of the ladder generates a torque that points into the page and has a magnitude equal to the product of the weight and its lever arm. The force exerted by the wall on the ladder generates a torque that points out of the page and has magnitude equal to the product of the supporting force and its lever arm.

$$\sum \tau_i = Mg \frac{L}{2} \sin\theta - F_{wall} L \cos\theta = MgL \left(\frac{1}{2} \sin\theta - \mu_S \cos\theta \right)$$

Setting the sum of these two torques equal to zero, we find that the tangent of the maximum angle the ladder can make with the wall before it starts to slip is just equal to twice the coefficient of static friction.

$$\sum \tau_i = 0 \qquad \Rightarrow \qquad \tan\theta = 2\mu_S$$

Let's now look at this result to see if the relationship between the angle and the coefficient of friction makes sense. Certainly, if there were no friction between the ladder and the ground ($\mu_S = 0$), then the ladder slips for all angles greater than zero. As the coefficient of friction increases, then the maximum angle also increases.

18.3   Example: Box on a Truck

Figure 18.2 shows a truck carrying a box of known length, width and mass which is uniformly distributed. The truck moves with constant acceleration and the frictional force is large enough to ensure that the box does not slip relative to the bed of the truck. If the acceleration is large enough, however, the box will tip over. Our job is to determine the maximum acceleration the truck can have before the box tips over.

We start by drawing a free-body diagram of the box as shown in the figure. The only horizontal force acting on the box is the frictional force exerted by the bed of the truck. Since the acceleration of the box is the same as that of the truck, the magnitude of this force must just be equal to the mass of the box times the acceleration of the truck.

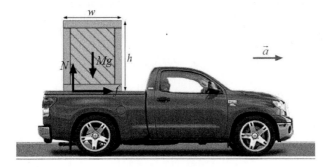

FIGURE 18.2 A truck, accelerating in the forward direction, carries a box of mass M, width w, and height h in its bed. The frictional force supplied by the bed of the truck on the box is sufficient to keep it from slipping.

$$f = Ma$$

Applying Newton's second law in the vertical direction determines the magnitude of the normal force on the box to be equal to the weight of the box.

$$N = Mg$$

We now know, for any given acceleration of the truck, the magnitudes of all the forces acting on the box. What determines the acceleration at which the box begins to tip? To answer this question, we need to consider the torques acting on the box. Since the box is accelerating, we must calculate the torques about an axis that pass through the center of mass of the box. Taking the direction of this axis to be perpendicular to the page, we can evaluate the torques produced by the three forces that act on the box.

The weight of the box produces no torque, since it acts at the center of mass of the box. What about the torques from the frictional and normal forces? To calculate these torques, we first need to determine where these forces act. When the box starts to tip, the only point of contact between the box and the truck will be at the lower back corner of the box Therefore, we calculate the torques produced by these forces at the time the box starts to tip by placing them at this single point of contact as shown in Figure 18.3. Consequently, we see that the frictional force produces a torque that points out of the page and has a magnitude equal to the product of the frictional force and its lever arm. Likewise, the normal force produces a torque that points into the page and has a magnitude equal to the product of the normal force and its lever arm.

$$\sum \tau_i = f(r \sin \theta) - N(r \sin \phi) = Ma_{max} \frac{h}{2} - Mg \frac{w}{2}$$

Setting the sum of these two torques equal to zero, we find that the maximum acceleration the truck can have without having the box tip over is just equal to the product of the ratio of the box's width to its height and the acceleration due to gravity.

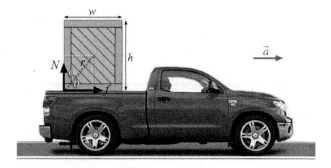

FIGURE 18.3 At the instant the box begins to tip, the bed of the truck exerts forces only on the lower back corner of the box.

$$a_{max} = g\frac{w}{h}$$

This result makes sense in that the box is harder to tip over to the extent that its width is large compared to its height!

18.4   Gravitational Potential Energy of Center of Mass

In the last unit we showed that the torque produced by the weight of an object around some rotation axis was proportional to the horizontal displacement of the center of mass of the object from the axis. One consequence of this result is that if the center of mass is directly below the axis, then the torque goes to zero and the object will not rotate; the object is in a state of static equilibrium. We will now show that this result can also be understood in terms of gravitational potential energy.

Figure 18.4 shows a beam attached to a wall at a pivot point. The object is free to rotate about an axis perpendicular to the page. What is the gravitational potential energy of this beam? Well, we know the gravitational potential energy of any mass element of the object (*dm*) is just equal to the product of its weight and its vertical displacement above some arbitrary height we have chosen to be the zero of potential energy. To find the

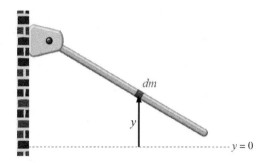

FIGURE 18.4 A beam is attached to a wall at a hinge. The beam's equilibrium position can be found either considering the torques about the hinge or the gravitational potential energy of the beam.

gravitational potential energy of the object, we just have to sum up the contributions from all mass elements that make up the object.

$$U_{gravity} = \int dU_{gravity} = \int dm\, gy$$

This integral is in fact proportional to the same integral that defines the position of the center of mass.

$$U_{gravity} = g\int y\, dm = gMY_{CM}$$

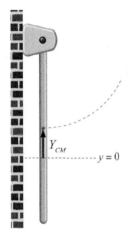

In particular, we see that the gravitational potential energy of the object is just equal to the product of the weight of the object and the vertical displacement of the center of mass from the height chosen to be the zero of potential energy.

Y_{CM}

$y = 0$

We have previously shown that an object tends to move in such a way that its potential energy is minimized. In this case, the potential energy is minimized when the center of mass of the object is as low as possible. Here, the center of mass is at its lowest point when the object hangs directly below the pivot as shown in Figure 18.5.

FIGURE 18.5 The gravitational potential energy is minimized when the beam hangs vertically.

Thus, we have shown that the position of equilibrium of a suspended object can be obtained either by considering the torque generated by its weight or by minimizing the gravitational potential energy!

18.5 Torque from the Gravitational Potential Energy

We have just shown that the position for stable equilibrium of an object can be obtained from considerations of either the net torque or the potential energy of that object. For example, we will consider the case that was shown in Figure 18.4. We will define the zero of potential energy to be the height of the center of mass when the center of mass is located directly below the pivot as shown in Figure 18.6.

If we now push the object to one side so that the line from the pivot to the center of mass makes an angle θ with the vertical as shown in Figure 18.7, the vertical displacement of the center of mass can be determined from trigonometry.

$$\delta y = \frac{L}{2}(1 - \cos\theta)$$

We can use this result to obtain the expression for the gravitational potential energy of the object as a function of θ.

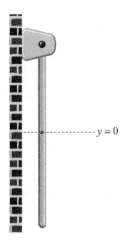

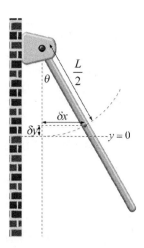

FIGURE 18.6 The   gravitational potential energy will be defined to be zero at the height of the center of mass.

FIGURE 18.7 The   gravitational potential energy is proportional to δy, the height of the center of mass of the beam above its equilibrium position.

$$U(\theta) = Mg\,\delta y = Mg\frac{L}{2}(1-\cos\theta)$$

We have previously shown that, for conservative forces, the force on an object in some direction is simply related to the change in the potential energy of the object in that direction.

$$\vec{F} = -\vec{\nabla}U$$

In this case, we can say that at any angle θ, the component of the gravitational force in the tangential direction is equal to minus the derivative of the gravitational potential energy with respect to the tangential direction.

$$F_{\text{tan}} = -\frac{dU}{ds_{\text{tan}}} = -\frac{dU}{d((L/2)\theta)} = -\frac{2}{L}\frac{dU}{d\theta}$$

The tangential element is equal to the product of the distance from the pivot to the center of mass and $d\theta$. Consequently, we see that the torque produced by the weight is equal to minus the derivative of the potential energy with respect to θ.

$$\tau = F_{\text{tan}}\frac{L}{2} = -\frac{dU}{d\theta}$$

Evaluating this derivative, we obtain

$$\tau = Mg\frac{L}{2}\sin\theta = Mg\,\delta x$$

where δx is the horizontal displacement of the center of mass from its equilibrium position as shown in Figure 18.7. Note that this result, which we obtained from a consideration of the change in the gravitational potential energy, is absolutely identical to the result that we obtained in the last unit in which we determined the torque directly from its definition!

18.6   Stability, Equilibrium, and the Center of Mass

We will close this unit with a brief discussion of the stability of objects. Figure 18.8 shows a trapezoidal structure sitting on a horizontal floor. To determine whether this object is stable or will fall over, we just need to find the location of the center of mass of the object relative to the footprint of the object on the floor.

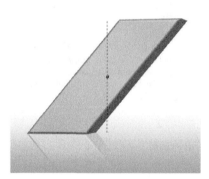

If the center of mass of the object is outside its footprint (as it is in Figure 18.8), then the object can lower its gravitational potential energy simply by falling down. If the center of mass is inside the footprint of the object (as it is in Figure 18.9), then it is stable.

We can verify this result by examining what happens to the height of the center of mass of the object if it is given a small push: the height of the center of mass increases a bit before eventually decreasing as the center of mass is pushed beyond the pivot point on the floor. This initial increase keeps the object from falling over on its own.

FIGURE 18.8 The stability of an object is determined by the location of its center of mass (green dot) relative to its footprint.

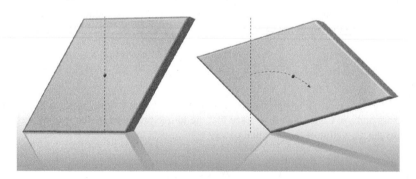

FIGURE 18.9 The object shown is stable. If given a small push, its center of mass initially increases, indicating an increase in gravitational potential energy.

MAIN POINTS

Gravitational Potential Energy

The gravitational potential energy of an extended object is equal to the product of the weight of the object and the vertical displacement of its center of mass from the height chosen to be the zero of potential energy.

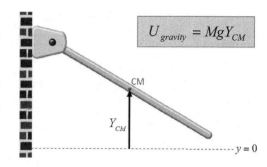

$$U_{gravity} = MgY_{CM}$$

Condition for Static Equilibrium

The position for static equilibrium of a suspended object can be determined in two equivalent ways:

1) The torque about the suspension axis is zero ($\tau_{Net} = 0$).

2) The gravitational potential energy is minimized ($dU_{gravity}/dy = 0$).

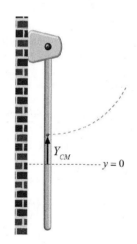

Condition for Stability

The condition for the stability of an extended object placed on a surface is that its center of mass must be located over its footprint on the surface.

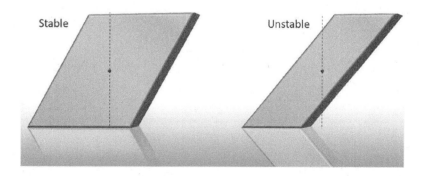

Stable   Unstable

PROBLEMS

1. Meterstick: A meterstick ($L = 1$ m) has a mass of $m = 0.23$ kg. Initially it hangs from two short strings: one at the 25 cm mark and one at the 75 cm mark. (a) What is the tension in the left string (i.e., the string at the 25 cm mark)? (b) The right string is cut! What is the initial angular acceleration of the meterstick about its pivot point? (You may assume the rod pivots about the left string, and the string remains vertical.) (c) What is the tension in the left string immediately after the right string is cut? (d) After the right string is cut, the meterstick swings down to where it is vertical for an instant before it swings back up in the other direction. What is the angular speed when the meterstick is vertical? (e) What is the acceleration of the center of mass of the meterstick when it is vertical? (f) What is the tension in the string when the meterstick is vertical? (g) When is the angular acceleration of the meterstick a maximum: *right after the string is cut and the meterstick is still horizontal, when the meterstick is vertical–at the bottom of its path*, or *the angular acceleration is constant?*

2. Ladder: A ladder leans against a smooth wall while also resting on a rough floor with a coefficient of static friction $\mu_s = 0.48$. The ladder has a length $L = 2.6$ m and a mass $m = 15$ kg. Assume the wall is frictionless. (a) What is the normal force the floor exerts on the ladder? (b) What is the minimum angle the ladder must make with the floor to not slip? (c) A person with mass $M = 65$ kg now stands at the very top of the ladder. What is the normal force the floor exerts on the ladder? (d) What is the minimum angle with respect to the floor to keep the ladder from sliding?

19

ANGULAR MOMENTUM

19.1 Overview

In this unit we will introduce a new vector quantity, the angular momentum, which plays the role for rotations that linear momentum plays for translations. In particular, we will use Newton's second law to obtain the important result that the sum of the external torques acting on a system about a specified axis is equal to the time rate of change of the angular momentum of the system about that same axis. We will first evaluate the angular momentum for several different systems and then we will apply this knowledge to determine the motion of systems in which the sum of the external torques is zero, which requires the angular momentum to be conserved.

19.2 Angular Momentum

In our study of rotations, we have defined quantities that are directly analogous to those we have developed for translational motion as shown in Table 19.1.

TABLE 19.1 Translational and rotational variable analogues.

Translational (Linear) Motion			Rotational Motion	
Displacement	$\vec{x}$	$\longleftrightarrow$	$\vec{\theta}$	Angular Displacement
Velocity	$\vec{v}$	$\longleftrightarrow$	$\vec{\omega}$	Angular Velocity
Acceleration	$\vec{a}$	$\longleftrightarrow$	$\vec{\alpha}$	Angular Acceleration
Mass	m	$\longleftrightarrow$	I	Moment of Inertia
Force	$\vec{F}$	$\longleftrightarrow$	$\vec{\tau}$	Torque

We can see this correspondence directly in the expressions for translational and rotational kinetic energy and in the equations that determine the motions.

$$\frac{1}{2}mv^2 \qquad \longleftrightarrow \qquad \frac{1}{2}I\omega^2$$

$$\vec{F}_{Net} = m\vec{a} \qquad \longleftrightarrow \qquad \vec{\tau}_{Net} = I\vec{\alpha}$$

We now want to introduce a new rotational quantity, the angular momentum, which corresponds to momentum. We know that it is a consequence of Newton's second law that if the sum of the external forces acting on a system is zero, then the total momentum of the system is conserved. We will define angular momentum in such a way that if the sum of the external torques acting on a system is zero, then the total angular momentum of the system will be conserved.

We start by defining the **angular momentum** $\vec{L}$ of a point particle about some axis as the cross product of $\vec{r}$, the vector from the axis to the particle, with the momentum vector as shown in Figure 19.1.

$$\vec{L} \equiv \vec{r} \times \vec{p}$$

Taking the derivative of $\vec{L}$ with respect to time, we get two terms.

$$\frac{d\vec{L}}{dt} = \left(\frac{d\vec{r}}{dt} \times \vec{p}\right) + \left(\vec{r} \times \frac{d\vec{p}}{dt}\right)$$

The first term is zero since the velocity vector ($d\vec{r}/dt$) and the momentum vector ($\vec{p}$) are parallel. The second term is just equal to the sum of the external torques on the particle ($\vec{\tau}_{Net} = \vec{r} \times \vec{F}_{Net} = \vec{r} \times d\vec{p}/dt$). Therefore,

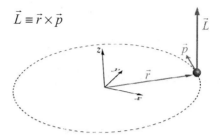

$$\vec{L} \equiv \vec{r} \times \vec{p}$$

FIGURE 19.1 The angular momentum of a point particle moving about an axis is defined as the cross product of the displacement vector and the momentum.

$$\frac{d\vec{L}}{dt} = \vec{\tau}_{Net}$$

This equation, then, identifies the time rate of change of the angular momentum as the sum of the torques that act on the particle. Defining the angular momentum of a system of particles as the vector sum of the angular momenta of the individual particles, we obtain our desired result: that the sum of the external torques acting on a system is equal to the time rate of change of the angular momentum of the system.

$$\vec{L}_{Total} \equiv \sum_i \vec{L}_i \qquad \Rightarrow \qquad \vec{\tau}_{Net} = \frac{d\vec{L}_{Total}}{dt}$$

Therefore, if the sum of these torques is zero, then the time rate of change of the angular momentum of the system is zero, which implies that the angular momentum of the system is conserved. For the remainder of this unit, we will look at some specific examples that should help to give some concrete meaning to this new concept of angular momentum.

19.3   Example: Point Particle Moving in a Circle

Figure 19.1 shows a particle of mass m moving with constant speed v in a circle of radius r around some rotation axis. To calculate the angular momentum of the particle as it goes around the axis we just need to evaluate the cross product of the displacement vector and the momentum ($\vec{L} \equiv \vec{r} \times \vec{p}$). The magnitude of this cross product is the product of r and p since the vectors are perpendicular. Using the right-hand rule we see that the direction of the angular momentum is along the $+z$ direction.

We can rewrite the magnitude of the angular momentum in terms of the angular velocity and the moment of inertia of the particle about that axis.

$$L = rp = rmv = \left(mr^2\right)\frac{v}{r} = I\omega$$

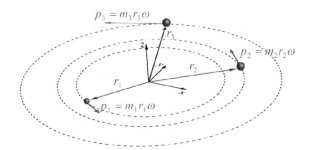

FIGURE 19.2 Three particles move with constant angular velocity ω about a common center. The angular momentum of the system, the sum of the angular momenta of the individual particles, is equal to the product of the moment of inertia of the system about the center and the angular velocity ω.

Indeed, if we instead consider a collection of point particles, all having different masses and distances from the rotation axis, as shown in Figure 19.2, we can determine the magnitude of the total angular momentum of the system by simply summing the magnitudes of the individual angular momentum, and we arrive at the same general expression we obtained for a single particle, except that the moment of inertia now refers to the moment of inertia of the system of particles about the axis.

$$L_{Total} = \sum_i L_i = \sum_i m_i r_i^2 \omega = I_{Total} \omega$$

This expression is very interesting. First, note its striking similarity to the expression for linear momentum. Namely, the mass has been replaced by the moment of inertia and the velocity has been replaced by the angular velocity. This correspondence is exactly the same as what we saw for translational and kinetic energy!

What about the direction of the angular momentum vector? Applying the right-hand rule for the angular momentum cross product for each particle, we see that the angular momentum of each particle, and therefore also the total angular momentum, points out of the page, if we view the motion from overhead. This direction is exactly the direction of the angular velocity vector. Recall that we defined the direction of this vector to be given by a right-hand rule. Namely, if you curl the fingers of your right hand in the direction of the rotation, then your thumb points in the direction of the angular velocity!

We therefore can summarize both the magnitude and direction of the total angular momentum by writing the simple vector equation

$$\vec{L} = I\vec{\omega}$$

where both $\vec{L}$ and $\vec{\omega}$ are vectors whose directions are found by using the right-hand rule.

19.4   Example: Solid Objects

We can extend our definition of the angular momentum to macroscopic solid objects by replacing the sum of angular momentum components over discrete particles to an integral over infinitesimal mass elements as we have done previously in our calculation of the moment of inertia of solid objects.

$$\vec{L} = \left(\sum_i m_i r_i^2 \right) \vec{\omega} \qquad \rightarrow \qquad \vec{L} = \left(\int dm\, r^2 \right) \vec{\omega}$$

In fact, this integral is exactly the integral that defines the moment of inertia of the object about the axis of rotation. Therefore, we obtain exactly the same equation for the angular momentum of a solid object as we have derived for a system of particles.

$$\vec{L} = I\vec{\omega}$$

For example, Figure 19.3 shows a solid disk spinning with a known angular velocity about its symmetry axis. What is the angular momentum of this disk? Since we already know the moment of inertia of a solid disk about its symmetry axis, the determination of the angular momentum is easy. The magnitude of the angular momentum is just the product of the moment of inertia and the angular velocity, and the direction of the angular momentum is the same as that of the angular velocity.

$$L = I\omega = \frac{1}{2} MR^2 \omega$$

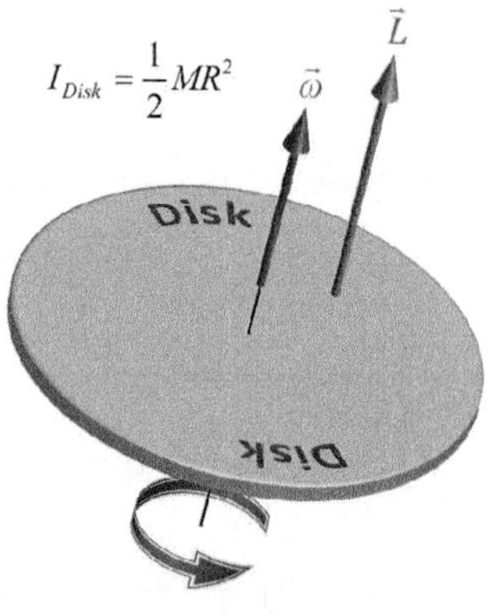

If the disk were replaced by a solid sphere, the expression for the angular momentum would look the same except for the numeric factor of one-half would become two fifths because of the change in moment of inertia. In all cases, the angular momentum is proportional to the angular velocity.

$$I_{Disk} = \frac{1}{2} MR^2$$

FIGURE 19.3 A disk spins with known angular velocity about its symmetry axis. The angular momentum has the direction of the angular velocity vector and a magnitude equal to the product of its moment of inertia and the angular velocity.

What if the spinning object were itself moving around some other rotation axis? A good example of such a situation is the Earth spinning around its own axis once per day while moving around the Sun once per year. The total angular momentum of the Earth about an axis through the Sun will be the sum of two distinct angular momentum components, one due to the motion of the center of mass of the Earth around the Sun, and one due to the rotation of the Earth about an axis through its own center of mass. Each of these two components is represented by a vector having a magnitude and a direction, and the total angular momentum of the Earth about the Sun will simply be the vector sum of these two components.

19.5   Example: A Disk Dropped on Another Disk

Our focus to this point has been the calculation of the angular momentum of a system. We will now illustrate the use of this concept in examples in which there are no external torques acting on the system. Figure 19.4 shows two uniform disks having the same radius and mass, one suspended above the other.

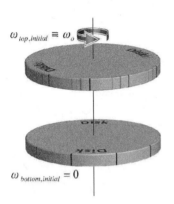

$\omega_{top,initial} \equiv \omega_o$

$\omega_{bottom,initial} = 0$

FIGURE 19.4 The top disk rotates about its axis of symmetry at constant angular velocity ω_o. It then drops and sticks to the bottom disk. The resulting angular velocity can be determined from the conservation of angular momentum.

Both are free to rotate about the same vertical axis. The top disk has an initial angular velocity ω_o and the bottom disk is initially at rest. The top disk is now dropped and lands on the bottom one. The frictional force exerted by the bottom disk on the top one opposes the motion and causes the top disk to slow down. The top disk exerts an equal and opposite force on the bottom disk, causing it to speed up. Eventually the disks reach the same angular velocity and the two disk system moves together with angular velocity ω_f. Our task is to determine this final angular velocity of the system.

The key here is to realize that torques that change the angular velocity of each disk are generated by the forces the disks exert on each other. If we consider our system to consist of both disks, then these torques will be *internal* torques. Since there are no external torques acting on the two-disk system, we know that the angular momentum of the system will be conserved. That is, the angular momentum of the system before the collision is equal to the angular momentum of the system after the collision.

The initial angular momentum is equal to the angular momentum of the top disk. The final angular momentum is equal to the sum of the angular momenta of each disk. Setting these expressions equal to each other, we determine that the final angular velocity of both disks is equal to one-half of the initial angular velocity of the top disk.

$$L_i = L_f \qquad \rightarrow \qquad \frac{1}{2}MR^2\omega_o = \frac{1}{2}MR^2\omega_f + \frac{1}{2}MR^2\omega_f$$

$$\omega_f = \frac{1}{2}\omega_o$$

This result is exactly analogous to the case of a one-dimensional totally inelastic collision between two equal mass blocks, in which one of the blocks is initially at rest.

We can take this analogy further by comparing the kinetic energies of the system before and after the collision. We can write the kinetic energy of the system in terms of the angular momentum, which is the same before and after, and the moment of inertia, which doubles.

$$K_i = \frac{L_i^2}{2I_{top}}$$

$$K_f = \frac{L_f^2}{2I_{both}} = \frac{L_i^2}{2(2I_{top})} = \frac{1}{2}K_i$$

We see that the final kinetic energy in this example is half of the initial kinetic energy, just as it was in the one dimensional totally inelastic collision.

19.6   Point Particle Moving in a Straight Line

To this point we have only discussed angular momentum in the context of rotating objects. Now the angular momentum of a particle about some given axis is defined, though, simply as the cross product of the vector from the axis to the particle and the momentum vector of the particle ($\vec{L} \equiv \vec{r} \times \vec{p}$). There is nothing in this definition that requires the particle to be rotating about that axis.

Figure 19.5 shows a particle moving with constant velocity past an axis which is perpendicular to the page. There are no forces acting on this particle; its momentum is clearly conserved. Since there are no forces acting on the particle, there are no torques acting on it either, so that its angular momentum must be conserved also.

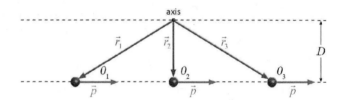

FIGURE 19.5 The angular momentum $\vec{L}$ of a particle moving with constant momentum $\vec{p}$ relative to the axis shown is given by the cross product of $\vec{r}$ and $\vec{p}$. $\vec{L}$ points out of the page, and its magnitude is a constant equal to the product of p and D.

If we now evaluate the magnitude of the angular momentum of the particle at any point along its path, we indeed see that we get the same value, namely the product of its momentum with its perpendicular distance to the axis.

$$L = pD$$

The direction of the angular momentum vector, at any point along the path, points out of the page. As this example illustrates, as long as there are no external torques acting on the system, the total angular momentum of a system is conserved even if the object is not moving in a circle. In the next section, we will solve a simple problem that involves both the angular momentum of rotation as well as the angular momentum from motion in a straight line.

19.7 Playground Example

Figure 19.6 shows a student running toward an empty merry-go-round at a playground. The student jumps onto its outer edge along a path that is tangent to the edge. How fast will the student and the merry-go-round end up spinning after the student has jumped on?

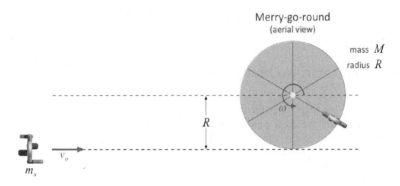

FIGURE 19.6 A student runs and jumps onto a merry-go-round.

To answer this question, we will define our system to consist of the student and the merry-go-round. We will choose the rotation axis to be the axle at the center of the merry-go-round, and we will assume that the axle is well greased so that we can ignore any friction when the merry-go-round rotates. With this assumption, we can say that there are *no external torques* acting on the system; therefore, the total angular momentum of the system must be conserved!

The initial angular momentum of the system about the axis of the merry-go-round is equal to the angular momentum of the student about the axis and is given by the product of the student's momentum and the perpendicular distance to the axis, which is equal to the radius of the merry-go-round.

$$L_i = m_s v_o R$$

After the student jumps on, both the student and the merry-go-round rotate with the same final angular velocity, so the final angular momentum of the system is equal to this angular velocity times the total moment of inertia of the system.

$$L_f = I_{Total}\omega$$

The total moment of inertia has two parts–one due to the merry-go-round and one due to the student. To determine this moment of inertia, we will take the merry-go-round to be a uniform disk of known mass and radius, and we will treat the student as a point particle at the edge of the merry-go-round.

$$I_{Total} = \left(\frac{1}{2} MR^2 + m_s R^2 \right) \omega$$

Since the angular momentum of the system is conserved, we can set the initial and final angular momenta equal to each other to obtain the final angular velocity.

$$L_i = L_f \qquad \rightarrow \qquad \omega = \frac{v_o}{R} \frac{2m_s}{M + 2m_s}$$

To get a feel for magnitudes here, we will take the student's mass to be 75 kg, the initial speed to be 2 m/s, and the mass and radius of the merry-go-round to be 150 kg and 2 m, respectively. For these numbers, we find that the final angular velocity is half a radian per second.

MAIN POINTS

Angular Momentum Definition

The angular momentum $\vec{L}$ of a point particle about some axis is defined to be the cross product of $\vec{r}$, the vector from the axis to the particle, with $\vec{p}$, the momentum vector of the particle.

$$\vec{L} \equiv \vec{r} \times \vec{p}$$

$$\vec{L}_{Total} = \sum \vec{L}_i$$

The angular momentum of a system of particles about a fixed axis is equal to the vector sum of the individual angular momenta.

The angular momentum for a system of particles rotating about a common axis with a fixed angular velocity is equal to the product of the moment of inertia of the system about the axis and the angular velocity vector.

$$\vec{L} = I_{system} \vec{\omega}$$

Rotational Equation of Motion

We used Newton's second law to obtain the rotational equation of motion, namely, that the sum of the external torques acting on a system is equal to the time rate of change of the angular momentum of the system.

$$\vec{\tau}_{Net,External} = \frac{d\vec{L}_{Total}}{dt}$$

Consequently, if the sum of the external torques on a system is zero, the angular momentum of the system is conserved.

PROBLEMS

1. Rod and Disk: A solid disk of mass $m_1 = 9.4$ kg and radius $R = 0.18$ m is rotating with a constant angular velocity of $\omega = 38$ rad/s. A thin rectangular rod with mass $m_2 = 3.9$ kg and length $L = 2R = 0.36$ m begins at rest above the disk and is dropped on the disk where it begins to spin with the disk. (a) What is the initial angular momentum of the rod and disk system? (b) What is the initial rotational energy of the rod and disk system? (c) What is the final angular velocity of the disk? (d) What is the final angular momentum of the rod and disk system? (e) What is the final rotational energy of the rod and disk system? (f) If rod took $t = 5.6$ s to accelerate to its final angular speed with the disk, what average torque was exerted on it by the disk?

2. Person on Disk: A person with mass $m_p = 75$ kg stands on the rim of a spinning platform disk. The disk has a radius of $R = 1.68$ m and mass $m_d = 193$ kg and is initially spinning at $\omega = 1.6$ rad/s. The person then walks two-thirds of the way toward the center of the disk (ending 0.56 m from the center). (a) What is the total moment of inertia of the system about the center of the disk when the person stands on the rim of the disk? (b) What is the total moment of inertia of the system about the center of the disk when the person stands at the final location two-thirds of the way toward the center of the disk? (c) What is the final angular velocity of the disk? (d) What is the change in the total kinetic energy of the person and disk? (A positive value means the energy increased.) (e) What is the centripetal acceleration of the person when the person is at $R/3$? (f) If the person now walks back to the rim of the disk, what is the final angular speed of the disk?

3. Merry-Go-Round: A merry-go-round with a radius of $R = 1.97$ m and moment of inertia $I = 193$ kg-m^2 is spinning with an initial angular speed of $\omega = 1.53$ rad/s. A person with mass $m = 57$ kg and velocity $v = 4.5$ m/s runs on a path tangent to the merry-go-round. Once at the merry-go-round she jumps on and holds on to the rim of the ride. (a) What is the magnitude of the initial angular momentum of the merry-go-round? (b) What is the magnitude of the initial angular momentum of the person 2 m before she jumps on the merry-go-round? (c) What is the magnitude of the initial angular momentum of the person just before she jumps on to the merry-go-round? (d) What is the angular speed of the merry-go-round after the person jumps on? (e) Once the merry-go-round travels at this new angular speed, what force does the person need to hold on? (f) Once the person gets half way around, she decides to simply let go of the merry-go-round to exit the ride. What is the linear velocity of the person right as she leaves the merry-go-round? (g) What is the angular speed of the merry-go-round after the person lets go?

4. Putty Rod (INTERACTIVE EXAMPLE): A piece of putty of mass $m = 0.75$ kg and velocity $v = 2.5$ m/s moves on a horizontal frictionless surface. It collides with and sticks to the end of a rod of mass $M = 2$ kg and length $L = 0.9$ m. The rod hangs vertically about a pivot at its other end. What fraction of the initial kinetic energy of the putty is lost in this collision, $K_{lost} / K_{initial}$?

20

ANGULAR MOMENTUM VECTOR AND PRECESSION

20.1 Overview

In this unit we will conclude our study of angular momentum by discussing several examples. We will start with examples in which there are no external torques acting on the system. In these cases, the angular momentum of the system, its magnitude and its direction, will be conserved. We will discuss the changes in angular velocities and kinetic energies that result from applying this conservation law. Finally, we will discuss an example in which the external torques acting on the system are not zero. Namely, we will discuss the gyroscope in which the weight of a spinning top provides a torque that changes the angular momentum of the system. We will find that the top precesses in the horizontal plane about its pivot at a rate determined by the torque about the pivot and the angular momentum about the axis of the top.

20.2 Example: Conservation of Angular Momentum

In the last unit we defined the angular momentum vector for a point particle about any axis ($\vec{L} \equiv \vec{r} \times \vec{p}$) and then used it to determine the expression for the angular momentum of a rigid rotating object ($\vec{L} = I\vec{\omega}$).

We determined that, just as linear momentum is changed by the application of an external force, *angular momentum* is changed by the application of an *external torque.*

$$\vec{\tau}_{Net,External} = \frac{d\vec{L}}{dt}$$

Consequently, if the net external torque on a system about some axis is zero, then the angular momentum of the system about that axis is conserved.

We will now do an example to illustrate these concepts. Figure 20.1 shows a particle, sliding without friction on a horizontal surface. The particle is constrained to move in a circle by a string. We can write the angular momentum of the particle about the rotation axis as the product of its moment of inertia ($I = mR^2$) and its angular velocity (ω).

$$L = mR^2\omega$$

We now turn on a little motor at the center of the circle that slowly winds in some string so that the radius of the circle is reduced from R to R_f. How will the angular velocity change, if at all?

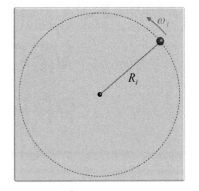

FIGURE 20.1 A particle slides without friction on a horizontal surface; it is constrained to move in a circle by a string.

The key here is to realize that there are no external torques acting about the rotation axis through the center of the circle. The torques generated by the normal force and the weight of the particle cancel, and the tension force generates no torque since its lever arm is zero.

Therefore, the angular momentum of the particle about the axis will be conserved as the string is shortened. During this process, the moment of inertia of the particle will decrease; therefore, its angular velocity must increase to keep the angular momentum constant. The final angular velocity increases by a factor of the ratio of the initial moment of inertia to the final moment of inertia.

$$\vec{L}_i = \vec{L}_f \qquad \rightarrow \qquad \omega_f = \frac{I_i}{I_f}\omega_i = \frac{R_i^2}{R_f^2}\omega_i$$

This analysis also explains why figure skaters, when they pull their arms and legs closer to their body, can spin faster since they have reduced their moment of inertia about the axis of rotation.

20.3   Angular Momentum and Kinetic Energy

In the last example, we saw that the angular velocity of the particle increased as the string was shortened. How do we understand this result in terms of energy?

We know the kinetic energy of the particle at any time is proportional to the square of its velocity. We can rewrite this expression in terms of the particle's angular momentum and moment of inertia about the rotation axis.

$$K = \frac{1}{2}mv^2 = \frac{1}{2}m\frac{L^2}{(mR)^2} = \frac{L^2}{2I}$$

Now, as the string was shortened, the particle's angular momentum remained constant, but its moment of inertia decreased. Therefore, its kinetic energy increased. In fact, the kinetic energy of the particle increases by a factor of the ratio of the initial moment of inertia to the final moment of inertia.

$$K_f = \frac{I_i}{I_f}K_i$$

Where did this kinetic energy come from? We know from the work-kinetic energy theorem that positive work must have been done on the particle in order to increase its kinetic energy. This work was done by the tension force! While the particle is being pulled inward, there is a component of its motion which is parallel to the tension and the dot product between force and displacement is non-zero. The total work done by the tension force must be equal to the change in the kinetic energy.

20.4   Vector Nature of Angular Momentum

To this point, the examples we have done to illustrate the conservation of angular momentum have only required consideration of the magnitude of the angular momentum. We will now consider an example in which we will need to explicitly deal with the vector nature of angular momentum.

Figure 20.2(a) shows a student sitting on a stool that can rotate freely in the horizontal plane. The student holds on to an axle that supports a top that spins without friction in the counterclockwise direction. Initially, the student and the stool are at rest and the top is spinning in the horizontal plane. The student now flips the top over as shown in

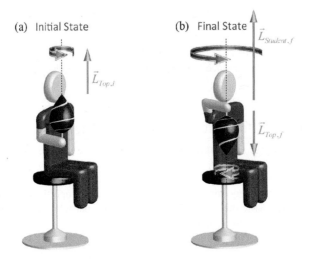

(a) Initial State   (b) Final State

FIGURE 20.2 A student sits on a stool that can freely rotate in the horizontal plane. (a) Initially he holds a top spinning so that its angular momentum vector points up. (b) He then flips the top over, causing himself and the top to rotate to conserve the total angular momentum of the system.

Figure 20.2(b), so that it is now upside down compared to its original orientation. What, if anything, happens to the student and the stool?

In this case, we will choose to define the system to be comprised of the student, the stool, and the spinning top. The only connection this system has to the external world is the bearing about which the stool turns. Since the stool can rotate freely, there are no external torques acting on the system about this rotation axis. Therefore, the total angular momentum of the system must be conserved!

Initially, the angular momentum of the system is just equal to the angular momentum of the spinning top. Once the top is flipped over, though, the student and the stool must start to rotate so that the total angular momentum of the system will be conserved. Namely, the angular momentum of the student plus the stool plus the angular momentum of the top after the flip must be equal, in both magnitude and direction, to the angular momentum of the top before the flip. Therefore, the student and stool end up rotating in the direction of the initial rotation of the top.

20.5 Precession

To this point, we have focused on systems in which the total external torque was zero and angular momentum was conserved. We will now turn to a class of problems where there is an external torque. This external torque will result in a change of angular momentum, often referred to as **precession**.

Figure 20.3(a) shows a disk spinning about its symmetry axis with an angular velocity ω. If the net external torque on this top is zero, then its angular momentum will be conserved, which means that it will keep spinning with the same speed and remain in the same orientation.

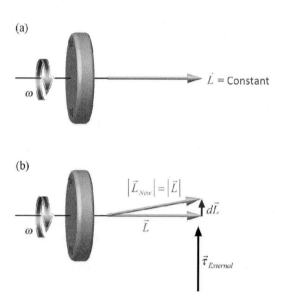

FIGURE 20.3 A disk spins with constant angular velocity about its symmetry axis. (a) In the absence of external torques, the angular momentum of the disk does not change in time. (b) If an external torque is applied, the angular momentum vector must change in the direction of the torque, giving rise to a precession of the spinning top.

In Figure 20.3(b), a torque is applied to the spinning disk in a direction that is perpendicular to the axis of the disk. Since this torque is perpendicular to the symmetry axis of the disk, it will not change the angular momentum of the disk about this axis. However, the equation that relates the net external torque to the time rate of change of the angular momentum requires the total angular momentum of the system to change. How is this possible?

The answer to this question comes from the equation itself–namely the direction in which the angular momentum changes is the same as the direction of the applied torque.

$$\vec{\tau}_{Net,External} = \frac{d\vec{L}}{dt}$$

The vectors representing the angular momentum of the disk about its axis and the external torque are shown in Figure 20.3(b). The vector representing the change in the angular momentum must be parallel to the net external torque vector. When we add this vector to our original vector, the direction of the new angular momentum vector changes, but its magnitude has not, since its magnitude must still be equal to the angular momentum of the spinning top about its axis. The axis of the disk has simply turned a bit in the direction of the applied torque. As long as the torque is being applied, the disk will continue to turn. We call this motion of the axis of the disk *precession*, and our next task will be to treat it quantitatively.

The vectors representing the angular momentum and the external torque applied to the spinning disk are shown in Figure 20.3(b). This torque acts for a small time dt, giving rise to a change in the angular momentum $d\vec{L}$ given by the product of external torque and dt. During this time interval, the angular momentum vector rotates through an angle $d\phi$, as shown in Figure 20.4. We can determine $d\phi$ using the small angle approximation. Namely, since dL is small compared to L, we can approximate dL as the product of L and $d\phi$.

FIGURE 20.4 The angular momentum vector $\vec{L}$ rotates through an angle $d\phi$ in a time dt. The change in the angular momentum dL is approximately equal to $L\, d\phi$.

$$dL \approx L\, d\phi$$

If we now differentiate the expression for the angular momentum with respect to time, we find that the magnitude of the time rate of change of the angular momentum is equal to the product of the angular momentum of the top and $d\phi/dt$, the precession rate of the top. This precession rate is often denoted by the capital Greek letter omega, Ω.

$$\frac{dL}{dt} \approx L\frac{d\phi}{dt} \equiv L\Omega$$

We know that the time rate of change of the angular momentum of a system is always equal to the net external torque acting on the system. We can rewrite our expression for the precession rate as the ratio of the net external torque applied to the system to the angular momentum of the system.

$$\Omega = \frac{\tau_{Net,External}}{L}$$

Ω has the same units as angular velocity–radians per second–and is exactly the angular velocity of the rotation of the axis of the disk as it is dragged around in a circle by the external torque.

At this point you probably have several questions: How could we actually apply such a torque that could cause this precession? What does this precession look like, really? In the next section we will work out a concrete example that should make this all clear.

20.6 Gyroscopes

Figure 20.5 shows a disk spinning about an axle through its symmetry axis. The axle is horizontal and is initially supported at both ends. In this configuration the net external torque on the disk is zero and the angular momentum of the disk is constant. In

Figure 20.6, the support on the left side is removed, and the remaining support acts as a frictionless pivot. The weight of the disk will now provide an external torque about this pivot. This force acts through a lever arm equal to the horizontal distance from the pivot to the center of the disk. Using the right-hand rule, we see the direction of this torque is perpendicular to both the axle through the disk and the weight of the disk. In other words, this torque will always be perpendicular to the angular momentum of the spinning disk about its rotation axis, just as we discussed in the previous section. Therefore, the disk will start to precess about the pivot.

Side View with Perspective
($\vec{g}$ points down the page)

FIGURE 20.5 A disk spins with constant angular velocity about its symmetry axis. The disk is supported at both ends.

What will this precession look like? Just as we saw in the last section, the angular momentum of the disk will turn in the same direction that the torque acts, as though it were being dragged around in a circle by the torque vector. This torque vector will always be parallel to the floor, so that the angular momentum vector will rotate in the horizontal plane. The rate of this precession is given by the ratio of the external torque produced by the weight of the disk to the angular momentum of the spinning disk.

$$\Omega = \frac{\tau_{Net,External}}{L} = \frac{Mgd}{L} = \frac{Mgd}{I\omega} = \frac{2gd}{R^2\omega}$$

This rate is typically quite small. For example, if the mass of the disk is 1 kg, its radius is 0.5 m, it is 0.5 m from the pivot, and it spins with an angular velocity of 100 rad/s, we calculate precession rate to be equal to 0.4 radians per second. In other words, it takes over sixteen seconds to make one complete revolution.

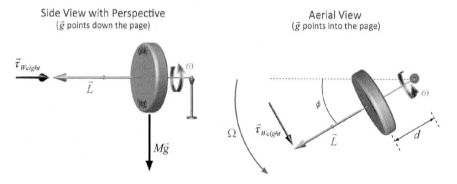

FIGURE 20.6 The spinning disk in Figure 20.5 has its left support removed. The weight of the disk exerts a torque on the disk that is perpendicular to the angular momentum vector, causing the disk to precess in the counterclockwise direction.

You may have noticed that our expression for omega blows up when the angular momentum of the disk approaches zero. Can we infer then that the precession rate goes to infinity as the angular momentum of the disk goes to zero? The answer is no, we can't. The problem is that the picture we have developed to derive this expression works well as long as the external torque is small compared to the angular momentum of the spinning object. Once this requirement is not met, the actual motion of the spinning disk becomes more complicated and its calculation is beyond the scope of this course.

MAIN POINTS

Conservation of Momentum Examples

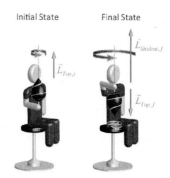

Initial State   Final State

The angular momentum of a system is conserved when there are no external torques acting on it.

Example 1: If a student sitting on a stool flips a spinning top over, he will begin to rotate in the direction of the initial orientation of the top.

Example 2: Both the angular velocity and the kinetic energy of a particle executing circular motion with a slowly decreasing radius will increase by a factor equal to the ratio of its initial to its final moment of inertia about the center.

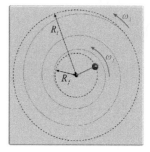

$$\omega_f = \frac{I_i}{I_f}\omega_i$$

$$K_f = \frac{I_i}{I_f}K_i$$

Precession Rate of a Gyroscope

The weight of a spinning disk provides a torque that changes the angular momentum of the system.

In particular, the torque is perpendicular to the angular momentum vector arising from the spinning of the disk, causing the disk to precess in the horizontal plane at a rate that is given by the ratio of the torque to the angular momentum due to the spin.

Aerial View
($\vec{g}$ points into the page)

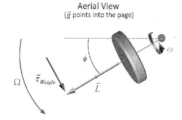

Precession Rate

$$\Omega = \frac{\tau_{weight}}{L} \qquad \text{where } \Omega \equiv \frac{d\phi}{dt}$$

PROBLEMS

1. Gyroscope: A disk with mass $m = 5.4$ kg and radius $R = 0.44$ m hangs from a rope attached to the ceiling. The disk spins on its axis at a distance $r = 1.24$ m from the rope and at a frequency $f = 18.9$ rev/s (with a direction shown by the arrow). (a) What is the magnitude of the angular momentum of the spinning disk? (b) What is the torque due to gravity on the disk? (c) What is the period of precession for this gyroscope? (d) What is the direction of the angular momentum of the spinning disk at the instant shown in the picture? *up, down, left,* or *right*. (e) What is the direction of the precession of the gyroscope: *it does not precess, clockwise as seen from above looking down the rope*, or *counterclockwise as seen from above looking down the rope*?

FIGURE 20.7 Problem 1

UNIT

21

SIMPLE HARMONIC MOTION

21.1 Overview

In this unit we will begin our study of oscillatory motion. Our primary example will be that of a mass attached to one end of a spring whose other end is fixed. This system oscillates about its equilibrium position with a frequency that is determined from the spring constant and the mass of the object. This motion is called simple harmonic motion. We will also offer an alternative description of this motion that makes clear its strong connection to uniform circular motion. We will determine the time dependence of the displacement, velocity, and acceleration of the object by applying Newton's second law to the system. We will find that all these quantities can be written in terms of sines and cosines whose argument is the product of the angular frequency of the oscillations and the time. Finally, we will demonstrate a procedure for obtaining the amplitude and phase of these oscillations.

21.2   Introduction to Oscillations

Figure 21.1 shows a mass that is free to move along a one-dimensional horizontal frictionless track. The mass is attached to one end of a spring, and the other end of the spring is fixed to the wall. We define $x = 0$ at the equilibrium position of the spring. If we displace the mass from the equilibrium position and release it, the mass oscillates about $x = 0$.

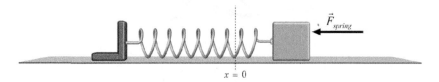

$x = 0$

FIGURE 21.1 A mass, resting on a horizontal frictionless surface, is connected to a support by a spring. If the mass is displaced from its equilibrium position and released, it will oscillate back and forth about the equilibrium position.

Since the spring force is the only force that does work on the mass, we know the total mechanical energy of the system will be conserved since the spring force is a conservative force. That is, the sum of the kinetic energy of the mass and the potential energy of the spring will be the same at any point of the motion. In particular, since the kinetic energy of the mass is zero at the endpoints of the motion, the potential energy of the spring will be maximized at those points. Similarly, since the potential energy of the spring is minimized at its equilibrium length, the kinetic energy of the mass will be maximized at this point.

We are now ready to study this motion in more detail. We know the acceleration of the mass will be maximized at the endpoints of the motion since the force the spring exerts on the mass will be a maximum there. We also know the acceleration of the mass will be zero at $x = 0$, since the force there is zero. The velocity of the mass will be maximum at $x = 0$ and will go to zero at the endpoints of the motion. As the mass oscillates, it continuously trades potential energy for kinetic energy.

In the next section we will use Newton's second law to build the quantitative description of this simple harmonic motion. Before moving on, though, we need to make one important point. Back when we first discussed the potential energy of a spring, we showed that the potential energy function for a mass on a vertical spring is identical to that for a mass on a horizontal spring as long as we measure displacements from the equilibrium length of the spring with the mass attached. Therefore, everything we have said thus far, and everything we are about to prove, will be equally true for both horizontal and vertical systems.

21.3 Simple Harmonic Motion

Figure 21.2 shows the horizontal mass plus spring system described in the last section at a time at which the mass is displaced a distance $\vec{x}$ from its equilibrium position.

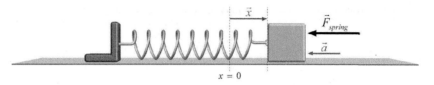

FIGURE 21.2 The mass is displaced a distance x from its equilibrium position. Newton's second law implies that the acceleration is in the direction of the force which points back toward the equilibrium position. The acceleration is not constant during this motion; it is proportional to the force which depends on the displacement $\vec{x}$.

We know the magnitude of the force is proportional to this displacement ($\vec{F}_{spring} = -k\vec{x}$). Further, from Newton's second law, we know the acceleration of the mass at this displacement.

$$\vec{F}_{Net} = m\vec{a} \qquad \rightarrow \qquad -k\vec{x} = m\vec{a}$$

What we do not know yet is the behavior of the system as a function of time, $x(t)$. The problem here is that the net force (and therefore the acceleration) is not a constant, so we cannot simply use our kinematic equations for constant acceleration. Instead, we will need to write Newton's second law as a differential equation and solve for the displacement as a function of time.

We can write Newton's second law explicitly in terms of the displacement by remembering that the acceleration is defined to be the time rate of change of the velocity and the velocity is defined to be the time rate of change of the displacement.

$$\vec{a} \equiv \frac{d\vec{v}}{dt} = \frac{d^2\vec{x}}{dt^2}$$

Therefore, Newton's second law for a mass on a spring can be written in terms of the displacement of the mass and the second time derivative of this displacement.

$$-kx = m\frac{d^2x}{dt^2} \qquad\qquad \text{(1-D motion)}$$

This equation is called a differential equation for the displacement. To solve it, we need to find a function $x(t)$, which when differentiated twice, gives back the negative of the same function $x(t)$ multiplied by a constant. It is convenient to define a quantity ω^2 as the ratio of the spring constant to the mass.

$$\omega^2 \equiv \frac{k}{m} \qquad \Rightarrow \qquad \frac{d^2x}{dt^2} = -\omega^2 x$$

We can solve this equation by inspection (that's the mathematician's word for guessing). Namely, we know that if you differentiate the sine of an angle you get the cosine of that angle, and differentiating once more gives you negative the sine of the angle. We get a similar result for differentiating the cosine of an angle twice. Therefore, we expect that our solution will involve sines and cosines. For example, we can directly demonstrate that $x(t) = A\sin(\omega t)$ and $x(t) = B\cos(\omega t)$ are both solutions to this differential equation:

$$x(t) = A\sin(\omega t) \qquad \Rightarrow \qquad \frac{d^2x}{dt^2} = -\omega^2 A\sin(\omega t) = -\omega^2 x$$

$$x(t) = B\cos(\omega t) \qquad \Rightarrow \qquad \frac{d^2x}{dt^2} = -\omega^2 B\cos(\omega t) = -\omega^2 x$$

Indeed, any function that is a linear combination of $\sin(\omega t)$ and $\cos(\omega t)$ will be a solution to our differential equation. Before discussing the physical meaning of this solution, we will make a slight digression to review some properties of sines and cosines in the next section.

21.4 Sines and Cosines

Figure 21.3 shows an oscillating sinusoidal function whose vertical axis represents the displacement of the oscillating mass and spring system as a function of time.

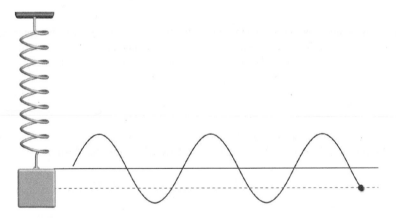

FIGURE 21.3 The vertical displacement of the mass on a spring as a function of time is represented by the sinusoidal function shown.

If you were asked "is this a sine or a cosine" all you would be able to answer is "yes." There is no difference in the overall form of a sine function and a cosine function–both oscillate symmetrically between +1 and −1 with a frequency determined by their argument, and both functions repeat after their arguments increase by 2π. It is not until we decide where to place $t = 0$ on the plot that the actual distinction between a sine or a cosine can be made. Indeed, if we start with a cosine function and we slide the $t = 0$ point to the left by $\pi/2$, it becomes a sine function.

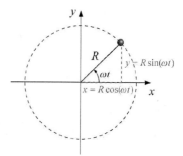

This connection between sines and cosines can also be understood geometrically. Figure 21.4 shows an object in uniform circular motion with radius R centered on the origin in the x-y plane. The angular displacement at any time t is given by the product of the time and the angular velocity. To find the x and y coordinates of the object at any time, we take the projections of the radius vector along the appropriate axis.

FIGURE 21.4 An object moves in a circle of radius R at a constant angular velocity ω. The projection of this motion along the x- and y-axes is identified as simple harmonic motion.

Consequently, we can understand the cosine and sine functions having identical shapes shifted by 90° relative to each other simply in terms of the 90° separation of the x- and y-axes.

This picture also explains why we use the term "angular velocity" to describe the oscillation frequency of an object executing simple harmonic motion. The displacement of the oscillating object from equilibrium is literally equivalent to the projection along a single axis of a particle moving in a circle with angular velocity ω. The two descriptions of the motion are just two sides of the same coin.

21.5 Example

We are now in a position to finish the example we introduced at the beginning of this unit. Figure 21.5 shows the mass given a displacement +D and released from rest at time $t = 0$. Our task is to determine the position of the mass as a function of time.

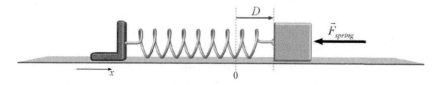

FIGURE 21.5 The mass is displaced a distance D from its equilibrium position and released from rest at time $t = 0$. The mass executes simple harmonic motion with a frequency determined by the spring constant and the mass.

We know the differential equation describing the oscillation of the mass has a general solution of the form

$$x(t) = A\cos(\omega t + \phi)$$

where $\omega^2 \equiv k/m$. We can determine the angular velocity ω since we know the spring constant and the mass. The remaining parameters (the **amplitude** A and the **phase angle** ϕ) are determined from the initial conditions. Since the mass was released from rest, we know, from energy conservation, that the displacement can never be larger than D. Consequently, the amplitude A is just equal to D. Further, since the initial displacement is equal to $+D$, we know that $\cos\phi$ must be equal to 1. Therefore, we take ϕ to be equal to zero.

$$x(t) = D\cos(\omega t)$$

We have now arrived at our solution: a function that determines the displacement as a function of time. We can obtain the velocity as a function of time by simply differentiating this function with respect to time.

$$v(t) = -\omega D\sin(\omega t)$$

One more differentiation gives the acceleration as a function of time.

$$a(t) = -\omega^2 D\cos(\omega t)$$

It is evident from these expressions that the velocity of the object is zero when its acceleration is a maximum (and vice versa). Note that the factor multiplying the sine or the cosine in these expressions for the velocity and acceleration is the maximum possible value for these quantities.

We will now plug in some numbers to get a feeling for these quantities. If we take the mass to be 2 kg, the spring constant to be 8 N/m and the initial displacement to be 0.5 m, we obtain an angular velocity of 2 radians per second, corresponding to a period of 3.14 seconds. In other words, it takes a bit over three seconds for the object to make one complete oscillation. Multiplying the angular velocity by the amplitude, we find the maximum velocity will be 1 m/s, while multiplying the square of the angular velocity by the amplitude, we find that the maximum acceleration will be 2 m/s^2.

21.6   Initial Conditions

In the last problem, we saw that the angular velocity (and therefore the period) of the motion was determined by the physical parameters of the system, the spring constant, and the mass. To complete the description of the motion, though, we also needed to determine the amplitude and the phase of the oscillations. These parameters were determined by the specification of the initial conditions, namely, that the mass was released from rest from an initial displacement. We chose those initial conditions in order to simplify the determination of the amplitude and the phase. We will now do another example with

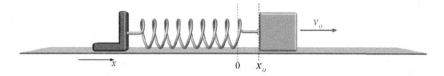

FIGURE 21.6 At time $t = 0$, the mass is displaced a distance x_o from its equilibrium position and moves with velocity $\vec{v}_o$. The amplitude and phase of the oscillations are determined by these two quantites.

different initial conditions to illustrate the general procedure for determining these quantities.

Figure 21.6 shows a mass oscillating on a spring, just as in the previous example, but we will not start the system from rest at a maximum displacement. Instead, we see that at $t = 0$, the displacement of the mass is x_o and its velocity is v_o.

How do we go about determining the amplitude and the phase of the oscillations? Since we know the displacement and the velocity at $t = 0$, we can just start from our general expressions for the displacement and velocity as a function of time and evaluate them at $t = 0$!

$$x(t) = A\cos(\omega t + \phi) \qquad \Rightarrow \qquad x(0) = x_o = A\cos(\phi)$$

$$v(t) = -\omega A\sin(\omega t + \phi) \qquad \Rightarrow \qquad v(0) = v_o = -\omega A\sin(\phi)$$

We now have two equations that we can solve simultaneously for the amplitude and the phase. For example, we can divide the first equation by the second to solve for ϕ.

$$\cot\phi = -\frac{\omega x_o}{v_o}$$

Knowing ϕ, we can use either of the two equations to solve for A.

$$A = \frac{x_o}{\cos\phi}$$

In this example, we demonstrated a general procedure for determining the amplitude and phase of the oscillations. Using these parameters we can obtain values for the displacement, velocity and acceleration of the oscillating mass at any time. Of course, many important parameters of the motion, for example the oscillation frequency and the energy in the system, are not time dependent and can be determined without any knowledge of the phase.

21.7   General Comments

One of the most important properties of simple harmonic motion in mechanics is that the frequency of the oscillations is determined totally by the constants needed to describe the net force that acts on the object. For example, in the spring-plus-mass system we have just studied, the frequency was determined from the spring constant and the mass.

Therefore, we see that the frequency of the oscillations does not depend on how the oscillations were started. For example, if we double the amplitude of the oscillations, thereby making the mass travel twice as far during every cycle, the time taken to complete that cycle is still exactly the same as it was when the amplitude was small!

This interesting property is a consequence of the fact that applying Newton's second law for this example results in the acceleration, the second derivative of the displacement with respect to time, being equal to the product of a negative constant and the displacement. Any time we find such an equation, we know we can write the solution in terms of sines and cosines with an oscillation frequency determined from the value of the negative constant. In our example the negative sign came about because the force was a restoring force; the force always acted in such a direction so as to restore the system to its equilibrium position. As we will show next time, other systems having a linear restoring force, such as a pendulum under certain conditions, will also undergo simple harmonic motion.

MAIN POINTS

Simple Harmonic Motion

A mass, resting on a horizontal frictionless surface connected to a spring, will oscillate about its equilibrium position with a frequency that is determined by the spring constant and the mass. This system is an example of simple harmonic motion.

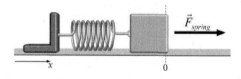

$$x(t) = A\cos(\omega t + \phi)$$
$$v(t) = -\omega A \sin(\omega t + \phi)$$
$$a(t) = -\omega^2 A \cos(\omega t + \phi)$$

The amplitude and the phase of these oscillations must be determined from the initial conditions, a specification of how the oscillation got started.

Angular Velocity (or Angular Frequency) $\omega = \sqrt{\dfrac{k}{m}}$

Uniform Circular Motion

Simple harmonic motion specified by an angular frequency ω and an amplitude A can also be represented as the projection of an object moving in a circular path of radius A with constant angular velocity ω along an axis that passes through the center of the circle.

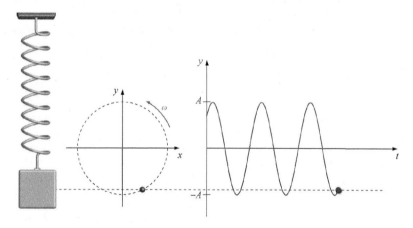

PROBLEMS

1. Vertical Spring: A block with mass $m = 6.5$ kg is hung from a vertical spring. When the mass hangs in equilibrium, the spring stretches $x = 0.26$ m. While at this equilibrium position, the mass is then given an initial push downward at $v = 4.5$ m/s. The block oscillates on the spring without friction. (a) What is the spring constant of the spring? (b) What is the oscillation frequency? (c) After $t = 0.32$ s what is the speed of the block? (d) What is the magnitude of the maximum acceleration of the block? (e) At $t = 0.32$ s what is the magnitude of the net force on the block? (f) Where is the potential energy of the system the greatest: *at the highest point of the oscillation, at the new equilibrium position of the oscillation,* or *at the lowest point of the oscillation?*

2. Mass on Two Springs: A block with mass $m = 6.2$ kg is attached to two springs with spring constants $k_{left} = 36$ N/m and $k_{right} = 50$ N/m.

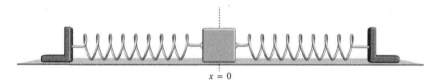

$x = 0$

FIGURE 21.7 Problem 2

The block is pulled a distance $x = 0.27$ m to the left of its equilibrium position and released from rest. (a) What is the magnitude of the net force on the block the moment it is released? (b) What is the effective spring constant of the two springs? (c) What is the period of oscillation of the block? (d) How long does it take the block to return to equilibrium for the first time? (e) What is the speed of the block as it passes through the equilibrium position? (f) What is the magnitude of the acceleration of the block as it passes through equilibrium? (g) Where is the block located, relative to equilibrium, at a time 0.96 s after it is released? If the block is left of equilibrium give the answer as a negative value; if the block is right of equilibrium give the answer as a positive value. (h) What is the net force on the block at time 0.96 s? A negative force is to the left; a positive force is to the right. (i) What is the total energy stored in the system? (j) If the block had been given an initial push, how would the period of oscillation change: *the period would increase, the period would decrease,* or *the period would not change?*

3. Block and Spring (INTERACTIVE EXAMPLE): At $t = 0$ a block with mass $M = 5$ kg moves with a velocity $v = 2$ m/s at position $x_o = -0.33$ m from the equilibrium position of the spring. The block is attached to a massless spring of spring constant $k = 61.2$ N/m and slides on a frictionless surface. At what time will the block next pass $x = 0$, the place where the spring is unstretched?

22

SIMPLE AND PHYSICAL PENDULA

22.1 Overview

In this unit, we will continue our study of oscillatory motion by studying a few specific examples. We will start by discussing the torsion pendulum, in which a disk suspended by a wire through its symmetry axis executes oscillations about that axis. We will find that the frequency of these oscillations is determined by the moment of inertia of the disk and the torsion constant of the wire. We will then investigate the physical pendulum in which an object hung from a frictionless pivot executes oscillations that are determined by the restoring torque that is provided by the weight of the object itself. We will find that these oscillations can be described as simple harmonic motion as long as the angular displacements are small. The frequency of these oscillations is independent of the mass of the object, being determined just by the size and shape of the object.

22.2 Simple Harmonic Motion

In the last unit we showed that an object that is subjected to a linear restoring force about some equilibrium position will oscillate around that equilibrium position, executing simple harmonic motion. The specific system we studied was that of a mass attached to a fixed spring, and we discovered that the oscillation was sinusoidal and the frequency of the oscillation did not depend on the amplitude of the oscillation, but only on the mass of the object and the stiffness of the spring.

Are there other systems that display similar behavior? The answer, of course, is yes. Simple harmonic motion is ubiquitous in all of nature and can be found everywhere from systems of atoms to systems of galaxies. We can identify a system that will exhibit such behavior if we can reduce the equations that govern its motion to the signature differential equation for simple harmonic motion that we derived last time.

$$\frac{d^2 x}{dt^2} = -\omega^2 x$$

In this unit we will once again use Newton's second law to identify several systems that exhibit simple harmonic motion.

22.3 The Torsion Pendulum

Figure 22.1 shows a disk suspended by a vertical wire attached to the center of the disk, parallel to its symmetry axis. The other end of the wire is attached to a fixed support. A rotation of the disk about its symmetry axis causes the wire to twist. Once twisted, the wire exerts a restoring torque on the disk. The magnitude of this torque is proportional to the angle through which the disk has been rotated.

$$\tau = -\kappa\theta$$

The direction of this torque is such that it causes the disk to return to its equilibrium position. If the disk is rotated through some initial angle and released, it will oscillate back and forth, executing simple harmonic motion.

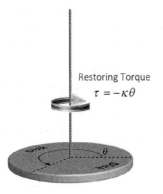

Restoring Torque

$$\tau = -\kappa\theta$$

FIGURE 22.1 A torsion pendulum: a disk is suspended by a vertical wire attached to the center of the disk, parallel to its symmetry axis. A rotation of the disk causes the wire to twist and exert a restoring torque to the disk. The disk then executes simple harmonic motion.

We can analyze this motion by applying Newton's second law for rotations. Writing the angular acceleration as the second derivative of the angular displacement, we obtain the signature differential equation for simple harmonic motion.

$$\alpha = \frac{\tau_{Net}}{I} \qquad \rightarrow \qquad \frac{d^2\theta}{dt^2} = \frac{-\kappa\theta}{I}$$

We identify the angular frequency of the motion as the square root of the torsion constant divided by the moment of inertia of the disk about its symmetry axis.

$$\omega^2 \equiv \frac{\kappa}{I}$$

We can now see from this equation of motion that the torsion pendulum is the rotational analog of the mass on a spring system. The roles that displacement, force, mass and the spring constant play for the mass on a spring system are played by the angular displacement, the torque, the moment of inertia and the torsion constant in the torsion pendulum. The general solution for the angular displacement of the pendulum as a function of time therefore has the same sinusoidal form as that of the displacement of the mass on a spring.

$$\theta(t) = \theta_{max} \cos(\omega t + \phi)$$

22.4 Pendula

Another class of systems that exhibit simple harmonic motion is that of pendula, objects that are supported from a pivot that have a restoring torque provided by gravity.

We have already seen that the weight of any object, hung from a frictionless pivot, exerts a torque that is proportional to the horizontal displacement of its center of mass away from the pivot. This torque is a restoring torque because it reduces this horizontal displacement. Figure 22.2 shows such a pendulum. We will now prove that this system exhibits simple harmonic motion as long as the angle that the pendulum makes with the vertical axis is not too big.

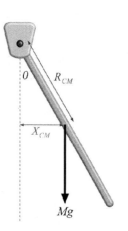

FIGURE 22.2 The pendulum: the weight of the rod exerts a restoring torque that causes the rod to swing back and forth. If the maximum angular displacement is small, the system exhibits simple harmonic motion.

We start with Newton's second law for rotations, and we evaluate both sides for motion around the pivot.

$$\tau_{Net} = I\alpha$$

On the left-hand side, the torque is equal to the weight of the pendulum times the horizontal displacement of the center of mass. The minus sign indicates that the torque opposes the displacement.

$$-MgX_{CM} = I\alpha$$

As long as the angle between the vertical direction and the line connecting the pivot to the center of mass isn't too big, we can approximate the horizontal displacement by the arc-length. With this approximation, we can write the displacement in terms of the angle and the distance from the pivot to the center of mass.

$$X_{CM} \approx R_{CM}\theta$$

The right-hand side of the equation is equal to the moment of inertia times the second time derivative of the angle, just as it was for the torsion pendulum in the previous section.

$$-MgR_{CM}\theta = I\alpha \qquad \rightarrow \qquad -\omega^2\theta = \frac{d^2\theta}{dt^2}$$

where $\omega^2 = MgR_{CM}/I$. We have now obtained the signature equation for simple harmonic motion with the angular frequency equal to the square root of the product of the weight of the pendulum and the distance of its center of mass from the pivot divided by its moment of inertia about the pivot. The variable that executes simple harmonic motion is the angle between the vertical direction and the line connecting the pivot to the center of mass.

In the next section, we will examine the simple pendulum consisting of a mass at the end of a string.

22.5   The Simple Pendulum

Figure 22.3 shows a simple pendulum, a point mass M attached to a massless string that is suspended from a pivot. We can apply our results from the last section with the distance from the pivot to the center of mass being equal to L, the length of the pendulum, and the

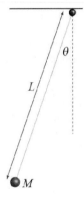

FIGURE 22.3 A simple pendulum: A point mass M is attached to a massless string that is attached to a pivot. If the maximum angular displacement is small, the system exhibits simple harmonic motion with a frequency determined by the length of the string.

moment of inertia equal to ML^2. The angular frequency of the motion then is equal to the square root of the acceleration due to gravity divided by the length of the pendulum.

$$\omega = \sqrt{\frac{g}{L}}$$

The period of the motion is, as always, given by 2π divided by the angular frequency and is therefore proportional to the square root of the length of the pendulum.

$$T = \frac{2\pi}{\omega} = 2\pi\sqrt{\frac{L}{g}}$$

To get a feeling for the magnitudes involved, we see that the period of a pendulum whose length is 2 meters will be a little less than 3 seconds. Now, two meters is a typical distance between you and the crossbar when you sit on a playground swing. Consequently, it makes sense that it takes you a few seconds to swing back and forth on a swing. Since the period doesn't depend on mass, it also makes sense that people having different masses can swing back and forth on a pair of playground swings while staying more or less side by side.

Just as was the case for the mass on a spring, we can use the initial conditions to determine the angle of the string as function of time.

$$\theta(t) = \theta_{max} \cos(\omega t + \phi)$$

For example, if we release the pendulum from rest, we know the amplitude will be given by the initial displacement, and the phase of the cosine function will be zero.

Before we leave this example, we want to point out a source of possible confusion using the symbol ω. The omega we have used in these equations is the angular frequency of the motion, a constant that does not depend on time. The angular velocity for the mass ($d\theta / dt$), which is often denoted also by ω, is *not a constant*. Indeed, $d\theta / dt$ itself oscillates with angular frequency ω!

22.6 Physical Pendula

We will now replace our simple pendulum by an object having a more complicated shape. Which properties of this oscillating system will be the same as those for a simple pendulum, and which will change? The answer is that the oscillating behavior is exactly the same as before; the only difference being that the expression for the frequency of the oscillation will look a bit different, depending on the location of the center of mass and the moment of inertia of the object.

For example, Figure 22.4 shows a stick of mass M and length L pivoted at one end. The distance from the pivot to the center of mass of the stick is half the length of the stick, while the moment of inertia of a thin rod about its end is given by $ML^2 / 3$. Plugging in

FIGURE 22.4 A physical pendulum: A stick of mass M and length L pivoted at one end. For small oscillations, the system executes simple harmonic motion with frequency which is equal to that of a simple pendulum whose lenth is two-thirds that of the stick.

these quantities, we find that this stick behaves exactly like a simple pendulum whose length is two thirds that of the stick.

$$\omega^2 = \frac{MgR_{CM}}{I} = \frac{Mg\dfrac{L}{2}}{\dfrac{1}{3}ML^2} = \frac{3}{2}\frac{g}{L}$$

Figure 22.5 shows another physical pendulum with a different shape and mass. We know we can find the period of this pendulum from its mass, its distance to the pivot and its moment of inertia. Since the moment of inertia will always be proportional to the mass of the object, we see that the period will be determined totally by the size and shape of the object.

FIGURE 22.5 A different physical pendulum. Its frequency is determined by its size and shape; the frequency is independent of its mass.

All of the results we have obtained so far for a pendulum oscillating due to gravity have depended on the assumption that the angle of oscillation is small. You may be wondering about this limitation. How small is small? As the angles increase, what happens to the oscillations? We will address these questions a bit more in the next section.

22.7   The Small Angle Approximation

We obtained the signature differential equation for simple harmonic motion for the angular displacement of a pendulum by assuming that the length of the arc swept out by the tip of a vector of length R, as it is displaced an angle theta from the vertical, is approximately equal to the horizontal distance moved by the tip of the vector (see Figure 22.6).

$$X_{CM} \approx R_{CM}\theta$$

The exact expression for this distance is proportional to the sine of the angle, not the angle itself.

$$X_{CM} = R_{CM}\sin\theta$$

Had we used this exact form, our differential equation would have looked almost the same, except we would have had $\sin\theta$ rather than θ on the right-hand side.

$$\frac{d^2\theta}{dt^2} = -\omega^2 \sin\theta$$

The up side of this equation is that it is an exact description of our system, valid for any angle. The down side of this equation, though, is that it does not have a simple analytic solution.

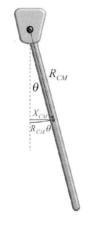

FIGURE 22.6 The difference between the arc length ($R_{CM}\theta$) and X_{CM} ($R_{CM}\sin\theta$).

To answer the question as to how small must the angle be so that the exact solution can be approximated by simple harmonic motion, we just need to look at how well θ approximates $\sin\theta$. We can make this comparison simply by using a calculator in its radian mode. Figure 22.7 shows a plot of the percentage difference between $\sin\theta$ and θ as a function of θ. We can see that the approximation is good to better than 5% for angles less than about 30°.

What happens to the motion of the pendulum as we increase the amplitude of the

Percent Difference Between θ and $\sin\theta$

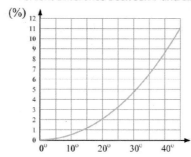

FIGURE 22.7 A plot of the percent difference between θ and $\sin\theta$ as a function of θ.

oscillations? The answer is not much, actually! The pendulum still has a fixed period, but a careful measurement of the angle as a function of time would show that the oscillations are no longer perfectly sinusoidal, and a careful investigation of the period of the oscillation would show that it is no longer completely independent of its amplitude. Consequently, we should be aware that the description of the motion of a pendulum as simple harmonic motion is not exact, but that it is, nonetheless, a very good description of the motion for moderate angles of oscillation.

MAIN POINTS

Torsion Pendulum

A disk, suspended by a wire through its symmetry axis, is an example of a torsion pendulum. The restoring torque provided by the wire is proportional to the angular displacement of the disk about its axis. The disk executes simple harmonic motion with a frequency proportional to the square root of the torsion constant divided by the moment of inertia about its symmetry axis.

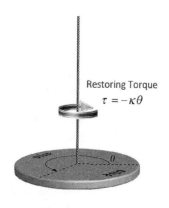

Restoring Torque
$$\tau = -\kappa\theta$$

Newton's 2nd Law for Rotations

$$\frac{d^2\theta}{dt^2} = -\omega^2\theta \qquad \text{where} \qquad \omega = \sqrt{\frac{\kappa}{I}}$$

Physical Pendulum

An object that executes oscillations, that are determined by the restoring torque provided by the weight of the object itself, is called a physical pendulum.

In the limit that the angular displacement from vertical is small, the motion can be described as simple harmonic motion with a frequency proportional to the square root of the weight of the object to its moment of inertia about the pivot times the distance between the pivot and the center of mass of the object.

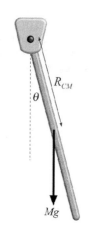

Newton's 2nd Law for Rotations

$$\frac{d^2\theta}{dt^2} = -\omega^2\theta \qquad \text{where} \qquad \omega = \sqrt{\frac{MgR_{CM}}{I}}$$

PROBLEMS

1. Simple Pendulum: A simple pendulum with mass $m = 1.7$ kg and length $L = 2.23$ m hangs from the ceiling. It is pulled back to a small angle of $\theta = 10.7°$ from the vertical and released at $t = 0$. (a) What is the period of oscillation? (b) At $t = 0$, what is the net force on the pendulum bob? (c) What is the maximum speed of the pendulum? (d) What is the angular displacement at $t = 3.45$ s? Give the answer as a negative angle if the angle is to the left of the vertical. (e) What is the magnitude of the tangential acceleration as the pendulum passes through the equilibrium position? (f) What is the magnitude of the radial acceleration as the pendulum passes through the equilibrium position? (g) Which of the following would change the frequency of oscillation of this simple pendulum: *increasing the mass, decreasing the initial angular displacement, increasing the length,* or *hanging the pendulum in an elevator accelerating downward*?

2. Torsion Pendulum: A torsion pendulum is made from a disk of mass $m = 6.8$ kg and radius $R = 0.76$ m. A force of $F = 46.2$ N exerted on the edge of the disk rotates the disk one-fourth of a revolution from equilibrium. (a) What is the torsion constant of this pendulum? (b) What is the minimum torque needed to rotate the pendulum a full revolution from equilibrium? (c) What is the angular frequency of oscillation of this torsion pendulum? (d) Which of the following would change the period of oscillation of this torsion pendulum: *increasing the mass, decreasing the initial angular displacement, replacing the disk with a sphere of equal mass and radius,* or *hanging the pendulum in an elevator accelerating downward*?

3. Physical Pendulum (INTERACTIVE EXAMPLE): A rigid rod of length $L = 1.0$ m and mass $M = 2.5$ kg is attached to a pivot mounted $d = 0.17$ m from one end. The rod can rotate in the vertical plane and is influenced by gravity. What is the period for small oscillations of the pendulum shown?

FIGURE 22.8 Problem 3

23

HARMONIC WAVES AND THE WAVE EQUATION

23.1 Overview

In this unit, we will begin our study of waves. We will start by discussing a specific type of wave, the transverse harmonic wave on a string, to introduce the general principles. In particular, we will develop the relationship between the speed of the wave and its frequency and wavelength. We will develop a mathematical expression that describes the displacement of the transverse harmonic wave as a function of space and time. We will then apply Newton's second law to a string under tension to obtain the equation that holds for any wave on a string. We will then demonstrate that our expression for the transverse harmonic wave is a solution of this equation with the velocity of the wave being determined totally by the tension in the string and its mass density. Finally, we will demonstrate that the harmonic wave on a string carries energy that is determined by the amplitude and frequency of the wave.

23.2   Harmonic Waves: Qualitative Description

In the last two units we have studied various systems that oscillate with simple harmonic motion. We will now consider systems made up of many objects that are coupled together and will study what happens when these systems are driven so that each part of the system undergoes simple harmonic motion.

Figure 23.1 shows a very long taut string lying on a horizontal frictionless floor. As we move one end of the string back and forth we see a pattern propagate down the string. Note that the motion of any point on the string is the same as that of the end of the string: It simply moves back and forth, perpendicular to the string. The only difference is that it reaches a maximum displacement somewhat later since it takes time for the pattern to propagate.

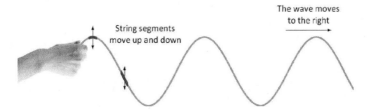

FIGURE 23.1 A wave is created in a taut string lying on a horizontal frictionless floor by moving one end back and forth. The motion of any part of the string (e.g., the highlighted pieces) is perpendicular to the direction of propagation of the wave. Indeed any piece of the string just oscillates about its equilibrium position. We call these waves transverse waves.

This simple picture captures the essence of the physics we will be studying for the rest of this unit as well as the next. The key observation is that even though there is clearly a pattern, or a wave, that keeps moving along the string to the right, the string itself is not going anywhere–any part of the string that we focus on just moves back and forth about a stationary equilibrium position. The string is merely the **medium** through which the wave travels. Indeed, we will soon see that the speed of the wave is determined totally by the characteristics of the medium.

The principles we will develop will apply to all kinds of waves. For example, consider the ever popular stadium wave often seen at sporting events. The wave travels around the stadium but the people that make up the wave do not; they simply stand up and sit down again as they follow the lead of the people next to them, just as one point on the string follows the lead of the point next to it. In the case of the stadium wave, the people are the medium in which the wave travels!

In the examples of the stadium wave and the string, the motion of any part of the medium at any given time is perpendicular to the direction in which the wave moves; consequently we say that these waves are **transverse**. When you study light waves in the second semester of this introductory physics sequence you will see that these waves are also transverse in nature. Waves do not have to be transverse, however. If you study the propagation of sound in a more advanced physics class you will see that even though the vibration of the medium in this case is actually parallel to the direction that the wave travels, all other aspects of the motion are identical to what you will learn in the next few sections.

23.3   Harmonic Waves: Quantitative Description

We will now quantify some characteristics of the wave we introduced in the last section. We'll start by creating the harmonic wave. Namely, we will move the end of the string just as a mass oscillates on a spring.

$$y_1(t) = A\cos(\omega t)$$

How must this expression change if we want it to represent the motion of a point some distance from the end of the string, for example at point 2 shown in Figure 23.2?

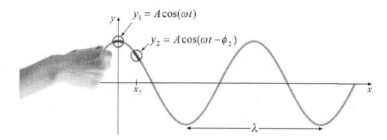

FIGURE 23.2 A harmonic wave is exited on a string by moving one end of the string as $A\cos(\omega t)$. The time dependence of a point (x_2) displaced from the end is different by a constant phase that depends on the distance. The distance between maxima (or minima) is called the wavelength λ of the wave.

The amplitude and frequency of the motion must be the same, but the actual displacement will lag behind that of the end of the rope. We can account for this difference by simply adding a negative term to the argument of the cosine function.

$$y_2(t) = A\cos(\omega t - \phi_2)$$

To check that this form works, just consider ϕ_2 to be the value of ωt when the point in question first reaches its maximum value. The argument of the cosine will be zero and the displacement will be a maximum at this time.

We will now consider a point twice as far from the end as the first point. Clearly, the negative constant we need to add to describe the motion of this point will be twice as big as that for the first point, since it takes twice as long for the wave to reach that point. Therefore, we can write an expression that works for all points on the string by setting $\phi = kx$, where x is the distance of the point from the end of the string and k is a constant called the **wave number**.

$$y(x,t) = A\cos(\omega t - kx)$$

If we freeze the wave at any given time we see that the shape of the wave repeats itself over and over as we move along the x-axis. The distance it takes for the pattern to repeat itself is called the **wavelength**, and we denote it by the Greek letter λ, as shown in Figure 23.2. Since a cosine repeats whenever its argument is increased by 2π, we can obtain the relationship between the wavelength and the wave number:

$$k\lambda = 2\pi \qquad \Rightarrow \qquad k = \frac{2\pi}{\lambda}$$

We now have obtained the general expression for the displacement of a transverse harmonic wave on a string traveling along the positive x direction. It is customary to swap the order of the two terms in the argument of the cosine so that the term containing x comes first.

$$y(x,t) = A\cos(kx - \omega t)$$

where $k = 2\pi / \lambda$.

23.4 Harmonic Waves

Figure 23.3 shows the displacement of a string as a function of time as a wave travels on it in the x direction. If we focus in on any specific piece of the string, we see that it is just moving from side to side with simple harmonic motion with frequency ω.

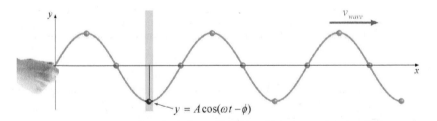

$$y = A\cos(\omega t - \phi)$$

FIGURE 23.3 A harmonic wave propagates in a taut string. The displacement of any piece of the spring oscillates with simple harmonic motion at the driving frequency.

The time it takes any piece of the string to make one complete oscillation is related to the frequency in exactly the same way we found when studying simple harmonic motion. We will use the symbol P to represent the period of the oscillation of the waves on a string, so that we can use T to represent the tension in the string.

$$P = \frac{2\pi}{\omega}$$

If we freeze this animation at $t = 0$ as shown in Figure 23.4, we see a harmonic wave along the x-axis whose amplitude and wavelength are easy to indentify. The wave number k is obtained simply from the wavelength ($k = 2\pi / \lambda$).

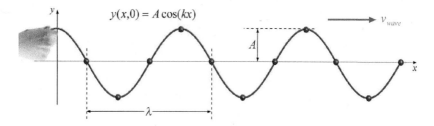

FIGURE 23.4 A snapshot of the string at $t = 0$. The wavenumber k is obtained from the wavelength λ ($k = 2\pi/\lambda$).

If we focus on any piece of the string, we see that exactly one wavelength passes through it as it makes one complete oscillation. In other words, the wave has moved one wavelength along the string during one period of its oscillation. The speed of the wave is therefore just the wavelength divided by the period.

$$v = \frac{\lambda}{P} = \frac{\omega}{k} = f\lambda$$

So what have we learned so far? We have made a plausible conceptual argument that if we drive the end of a string from side to side with simple harmonic motion, then all points in the string will eventually be moving with simple harmonic motion as the wave propagates along the string. We have obtained an expression for the displacement from equilibrium of any part of the string at any time, and we have seen how the speed of the wave is related to its period and the wavelength.

In the next two sections we will apply Newton's second law to obtain this exact form and to determine how the speed of the wave depends on the tension and the mass density of the string.

23.5 Newton's Second Law

Figure 23.5 shows a small element dm of the string to which we will apply Newton's second law to obtain the wave equation. We will assume the tension in the string is the same everywhere along its length, and the displacement of any part of the string from equilibrium is small compared to the wavelength. This second assumption enforces the requirement that the angle that any part of the string makes with the x-axis will always be small, which will simplify our calculation.

In order to apply Newton's second law, we will need to determine the net force on a small element of string. The net force in the x direction is the difference in the x components of the tension on either side of the string element. Since the cosine of a small angle is very

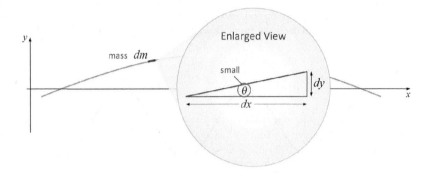

FIGURE 23.5 In order to develop the wave equation from Newton's second law, we assume the tension in the string is the same everywhere and that the displacement of any part of the string from its equilibrium position is small with respect to its wavelength. With these assumptions, we can use the small angle approximation shown: $\theta \approx \sin\theta \approx \tan\theta = dy/dx$.

close to one, this difference will be close to zero and the net force on the string element in the x direction vanishes.

$$\sum F_x = T(\cos\theta_2 - \cos\theta_1) \approx 0$$

The net force in the y direction is the difference in the y components of the tension on either side of the string element. Since the sine of a small angle is very close to the angle itself, this difference will be just the tension times the difference in the angle on either side of the string element.

$$\sum F_y = T(\sin\theta_2 - \sin\theta_1) \approx T(\theta_2 - \theta_1) = T\,d\theta$$

We can now apply Newton's second law to the string element in the y direction. We know the net force, and we will write the mass of the element as the product of the length of the string and μ, its mass density. The resulting expression then relates the product of the tension and the derivative of the angle of the string with respect to x to the product of the mass per unit length and the acceleration.

$$T\,d\theta = dm\,a_y \qquad \rightarrow \qquad T\,d\theta = (\mu\,dx)a_y \qquad \Rightarrow \qquad T\frac{d\theta}{dx} = \mu a_y$$

Since we want to end up with an expression for the displacement of y as a function of x and time, we want to eliminate the angle θ from this expression. Now $\tan\theta$ is just equal to dy/dx. Since we have assumed θ to be small, we can approximate $\tan\theta$ as just θ. Therefore, we can replace $d\theta/dx$ in our expression by the second derivative of y with respect to x.

$$T\frac{d^2y}{dx^2} = \mu a_y$$

The left-hand side of this equation is proportional to the net force acting on an element of the string. Therefore, we see that the net force on an element of string is proportional to the curvature d^2y/dx^2 of the string. The net force then acts as a restoring force since it always points back toward the equilibrium position ($y = 0$).

The acceleration can always be written as the second derivative of the displacement with respect to time. When we make this substitution into our equation that we obtained from Newton's second law, we arrive at a general equation that holds for all waves that can be supported in a string.

$$\frac{d^2y}{dx^2} = \frac{\mu}{T}\frac{d^2y}{dt^2}$$

We will now verify that the form we previously obtained for a harmonic wave is a solution to this general equation! Namely, we assume

$$y(x,t) = A\cos(kx - \omega t)$$

and differentiate twice with respect to position to obtain

$$\frac{d^2y}{dx^2} = -k^2\left[A\cos(kx - \omega t)\right] = -k^2y$$

We now differentiate twice with respect to time to obtain

$$\frac{d^2y}{dt^2} = -\omega^2\left[A\cos(kx - \omega t)\right] = -\omega^2y$$

Therefore, we see that our form for the harmonic wave is indeed a solution of the equation as long as the ratio of ω to k is equal to the square root of the tension divided by the mass per unit length of the string.

$$\frac{\omega}{k} = \sqrt{\frac{T}{\mu}}$$

We know the ratio of ω to k is the same as the ratio of λ to P, which is equal to the velocity of the wave! Therefore, we obtained the important result that the velocity of the wave on a string is determined totally by the tension in the string and its mass per unit length.

$$v = \sqrt{\frac{T}{\mu}}$$

The velocity is indeed determined by the medium; it does not depend on the frequency, the wavelength, or the amplitude of the wave itself. For example, if we double the frequency of a wave, its wavelength will decrease by a factor of two, keeping the speed of the wave the same!

23.6   The Wave Equation

We just obtained the important result that the speed of waves on a string depends only on properties of the string itself, namely its tension and its mass per unit length. This result follows from the fact that Newton's second law in this case produces an equation known as the **wave equation** that states that the second derivative of the displacement with respect to space is proportional to the second derivative of the displacement with respect to time, where the constant of proportionality is given by one over the speed of the wave squared.

$$\frac{d^2 y}{dx^2} = \frac{1}{v^2} \frac{d^2 y}{dt^2}$$

This equation is the defining equation for waves. The characteristic speed of the wave is determined by the constant in this equation. This constant will depend only on the properties of the medium for the wave. For example, for sound waves, the medium is the air, and the speed depends on factors like the air pressure and temperature, as well as the properties of the molecules in the air.

The fact that the speed of the wave does not depend on the wavelength or frequency of the wave should be consistent with your experience. For example, you know that sound waves of all frequencies must travel at about the same speed since a person talking to you from down the hall sounds about the same as he would standing right next to you–the sounds from a distance will be quieter but the voice, which is made up of both high and low frequencies, will still sound the same!

23.7   Waves Carry Energy

We'll close this unit by considering one more important quality possessed by waves–their ability to transmit energy from one place to another. Clearly, energy is needed to start the wave. If we consider the wave on a string as an example, we know the string will possess kinetic energy since it has mass and the passing wave causes its elements to move. Indeed, we can see how this energy will scale with the frequency and amplitude of the wave by obtaining an expression for the velocity of an element of the string by differentiating, with respect to time, our expression for its displacement.

$$v_y(x,t) \equiv \frac{dy}{dt} = \frac{d}{dt}\left[A\cos(kx - \omega t)\right] = \omega A \sin(kx - \omega t)$$

Just as we found when studying simple harmonic motion, we see that the maximum velocity of an element of the string is equal to the product of the frequency and amplitude of the wave.

$$v_{max} = \omega A$$

Therefore, the maximum kinetic energy of any element of the string is proportional to the square of this product. This result is, in fact, general, holding for all kinds of waves.

$$K_{max} \propto \omega^2 A^2$$

We have just determined the scaling behavior for the kinetic energy of any piece of our string. We now will demonstrate that energy moves down the string in the same direction as that of the wave. The key here is to realize that each element of the medium is driven by the one just before it. Consider the hand pumping the end of the string from side to side as shown in Figure 23.3. As the hand moves, it exerts a force on the string in the direction of its motion. Therefore, the hand always does positive work on the string, causing the kinetic energy of the string to increase. We see this increase of energy since more and more of the string moves as the wave propagates down the string. As the wave moves down the string with a speed v, so does the energy!

So far we have considered only harmonic waves in which the particles in the medium oscillate with simple harmonic motion and as a result the waves themselves can be represented by a sinusoidal function of space and time. In the next unit, we will show that all waves are *not* represented by sinusoidal functions. Indeed, a function of just about any shape at all will satisfy the wave equation as long as it represents motion with a constant speed.

MAIN POINTS

Transverse Harmonic Waves

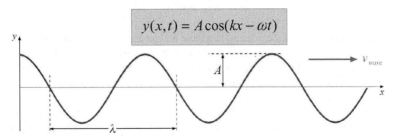

$$y(x,t) = A\cos(kx - \omega t)$$

The displacement of an element of the medium through which a transverse harmonic wave propagates is described as a sinusoidal function of space and time.

Period of Oscillation $\quad P = \dfrac{2\pi}{\omega}$

Wave Number $\quad k = \dfrac{2\pi}{\lambda}$

The speed of such a wave is given by the product of its wavelength and its frequency.

Speed of Wave $\quad v_{wave} = \dfrac{\lambda}{P} = \dfrac{\omega}{k} = f\lambda$

The Wave Equation

Applying Newton's second law to a string under tension yields an equation that holds for any wave on a string.

$$\frac{d^2 y}{dx^2} = \frac{1}{v^2}\frac{d^2 y}{dt^2}$$

The velocity of the wave is identified as the square root of the tension in the string divided by its mass density.

$$v_{wave} = \sqrt{\frac{T}{\mu}}$$

Energy in Waves

Energy propagates with the wave and its maximum value is proportional to the square of the product of its amplitude and its frequency.

Max Kinetic Energy of a String Element

$$K_{max} \propto \omega^2 A^2$$

PROBLEMS

1. Wave Equation: A transverse harmonic wave travels on a rope according to the following expression: $y(x,t) = 0.16\sin(2.2x + 17.3t)$. The mass density of the rope is $\mu = 0.129$ kg/m. x and y are measured in meters and t in seconds. (a) What is the amplitude of the wave? (b) What is the frequency of oscillation of the wave? (c) What is the wavelength of the wave? (d) What is the speed of the wave? (e) What is the tension in the rope? (f) At $x = 3.8$ m and $t = 0.47$ s, what is the velocity of the rope? (g) At $x = 3.8$ m and $t = 0.47$ s, what is the acceleration of the rope? (h) What is the average speed of the rope during one complete oscillation of the rope? (i) In what direction is the wave traveling: $+x$ *direction, $-x$ direction, $+y$ direction, $-y$ direction, $+z$ direction, or $-z$ direction?* (j) On the same rope, how would increasing the wavelength of the wave change the period of oscillation: *the period would increase, the period would decrease,* or *the period would not change?*

2. Wave Pulse: A wave pulse travels down a slinky. The mass of the slinky is $m = 0.9$ kg and is initially stretched to a length $L = 7.8$ m. The wave pulse has an amplitude of $A = 0.24$ m and takes $t = 0.46$ s to travel down the stretched length of the slinky. The frequency of the wave pulse is $f = 0.49$ Hz. (a) What is the speed of the wave pulse? (b) What is the tension in the slinky? (c) What is the average speed of a piece of the slinky as a complete wave pulse passes? (d) What is the wavelength of the wave pulse? (e) The slinky now is stretched to twice its length (but the total mass does not change). What is the new tension in the slinky? Assume the slinky acts as a spring that obeys Hooke's Law. (f) What is the new mass density of the slinky? (g) What is the new time it takes for a wave pulse to travel down the slinky? (h) If the new wave pulse has the same frequency, what is the new wavelength? (i) What does the energy of the wave pulse depend on: *the frequency, the amplitude,* or *both the frequency and the amplitude?*

UNIT

24

WAVES AND SUPERPOSITION

24.1 Overview

In this unit, we will conclude our study of waves. We will begin by demonstrating that any function whose argument is proportional to $x - vt$ is a solution to the wave equation with wave velocity equal to v. We will then clarify that the equation that defines mechanical waves must be derived from Newton's second law and that the wave velocity will be indentified from the constants that specify the forces and masses in the problem. We will then apply the principle of superposition to demonstrate the existence of standing wave solutions to the wave equation that are generated by two waves of equal amplitude and frequency that move with opposite velocities on the same string. Finally, we will show how such standing waves are generated when an incident wave is reflected from a fixed endpoint.

24.2 General Solution to the Wave Equation

We began our study of waves in the last unit. We saw that as a wave propagates through a medium, the elements of the medium oscillate about their equilibrium position while the wave can propagate indefinitely. We studied the example of a wave on a string, and we derived the wave equation–a differential equation whose solution describes the displacements of the medium as a function of both position and time.

$$\frac{d^2 y}{dx^2} = \frac{1}{v^2} \frac{d^2 y}{dt^2}$$

We demonstrated that one such solution was the harmonic wave, with the speed of the wave being given by the ratio of the angular frequency to the wave number. We will now show that waves of any shape–not just harmonic waves–will satisfy this equation as long as the speed of the wave is constant.

We start by rewriting the argument of the cosine in our harmonic solution in terms of the velocity of the wave. We see that the argument of the cosine is now proportional to the quantity $x - vt$.

$$kx - \omega t = k(x - \frac{\omega}{k} t) = k(x - vt)$$

We will now show that any function whose argument is proportional to $x - vt$ is a solution to the wave equation.

We use the chain rule to differentiate our function with respect to both x and t.

$$y = f(x - vt) \qquad \Rightarrow \qquad \frac{dy}{dx} = f' \qquad \Rightarrow \qquad \frac{d^2 y}{dx^2} = f''$$

$$y = f(x - vt) \qquad \Rightarrow \qquad \frac{dy}{dt} = -v f' \qquad \Rightarrow \qquad \frac{d^2 y}{dx^2} = (-v)^2 f''$$

Each time we differentiate with respect to t, we bring out a factor of $-v$. Consequently, we see that the second derivative of our function with respect to t is equal to v^2 times the second derivative of our function with respect to x.

$$\frac{d^2 y}{dx^2} = \frac{1}{v^2} \frac{d^2 y}{dt^2}$$

Note that our function $f(x - vt)$ does not have to be harmonic at all. It can represent a wave of any shape moving with speed v. For example, Figure 24.1(a) shows a shape described by the function $y = f(x)$ that is centered at the origin.

Figure 24.1(b) shows the shape moved to the right a distance d along the x-axis. The new shape is described by exactly the same function if we just change the argument from x to $x - d$, since the new argument will have the same value when $x = d$ as the old one did when $x = 0$. If this shape (we'll call it a pulse) moves to the right with constant speed v, we can describe the motion completely by simply replacing "d" by "vt" in the argument of the function!

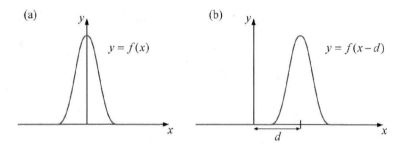

FIGURE 24.1 (a) A pulse described by $y = f(x)$ is centered at the origin. (b) The same pulse is displaced a distance d to the right and is now described as $f(x - d)$. If the pulse in (a) moves to the right with constant speed v, we describe the motion as $f(x - vt)$!

24.3   The Speed of a Wave is Determined by the Medium

We have just shown that any function whose argument is equal to $x - vt$ is a solution to the wave equation for a wave having speed v. Note that the speed v is defined in the wave equation. For a wave on a string, we used Newton's second law to obtain a wave equation described by a speed which is equal to the square root of the tension divided by the mass density of the string.

$$v = \sqrt{\frac{T}{\mu}}$$

The important point we want to make here is that the speed of the wave is determined by the medium and does not depend on how the wave was initially excited!

For example, Figure 24.2(a) shows the wave generated by our initial harmonic excitation, while Figure 24.2(b) shows a single pulse initiated in the string. Note that the speeds of these two different waves are the same, as they must be, since the speed is totally determined by the medium. In both cases, the descriptions are given as functions of "$x - vt$," as the wave is traveling in the positive x direction.

We can also excite a wave in the string by driving the opposite end of the string. In this case, the wave travels to the left and is represented by the same function, except that we just have to change the sign of the velocity; that is, the description of a wave moving in the negative x direction is given as a function of "$x + vt$."

(a)

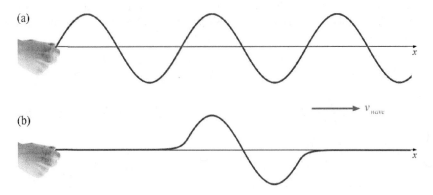

V_{wave}

(b)

FIGURE 24.2 (a) A continuous wave is excited in the string. The wave pattern moves with velocity $\bar{v}$ to the right. (b) A single pulse is excited in the string. The pulse also moves with velocity $\bar{v}$ to the right.

24.4 Superposition

We have just shown that any function of the form $f(x \pm vt)$ is a solution to the wave equation where v, the speed of the wave, is determined by the properties of the medium. If two different functions are separate solutions to the wave equation, then their sum is also a solution. At first, this observation may not seem to be significant, but, in fact, it has profound implications for certain properties of waves. Namely, if many waves can travel in the same medium at the same time, each wave will move independently of the others, resulting in a total displacement of the medium at any time and place that is simply equal to the *sum* of the individual displacements.

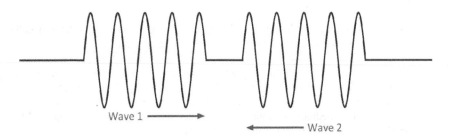

Wave 1 ⟶

⟵ Wave 2

FIGURE 24.3 Two harmonic waves of equal amplitude and wavelength move in the same string. As they pass through each other, the displacement at any particular point on the string is equal to the sum of the displacements from wave 1 and wave 2.

Figure 24.3 shows an important example of this result. Two harmonic waves having the same amplitude and wavelength move along a string in opposite directions. As they meet, the total displacement at that position and time is given by the sum of the individual displacements as shown in Figure 24.4.

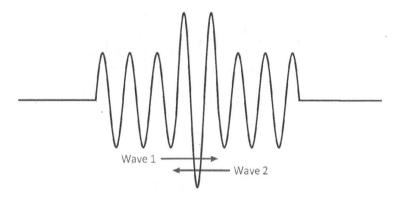

FIGURE 24.4 The superposition of the two traveling waves shown in Figure 24.3 at a time shortly after the leading edges meet.

There are two important things to notice here. First, as the waves pass each other, the sum seems to be standing still. Second, after the waves pass each other they both look exactly the same as they did before they met.

Both of these observations are interesting. The fact that the waves look the same before and after they pass through each other means that the waves do not interact; one wave does not in any way change the nature of the other. This simple result has important consequences. For example, this non-interaction of the waves allows many phone calls to be carried by a single optical fiber at the same time and insures that the light from a distant star can reach a telescope on Earth.

The fact that the sum of the two waves appears to be standing still is also extremely important. This phenomenon, called **standing waves**, allows us to build both microwave ovens and guitars, for example. We will investigate the properties of standing waves in more detail in the next section.

24.5   Standing Waves

We just made the claim that when two waves having the same amplitude and wavelength but traveling in opposite directions meet the result is a wave which stands still. Figure 24.5 shows snapshots of the string at three time intervals as the traveling waves pass through each other. The first feature to notice is that the amplitude of the standing wave goes from a minimum of zero when the displacements of the incoming waves cancel, to a maximum when the incoming waves lay directly on top of each other. This amplitude keeps oscillating as the incoming waves move past each other. Second, we see that the wavelength of the standing wave is the same as that of the incoming waves. Third, we see that there are points along the string whose displacement is always zero!

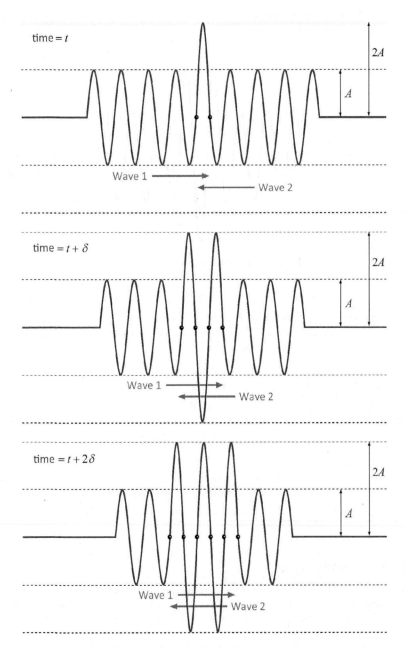

FIGURE 24.5 Three snapshots of the traveling waves from Figure 24.3 as they pass through each other, forming a standing wave. Note that some locations along the string, marked with black dots, are **nodes** where the displacement is always zero. Another important feature is that the apparent wavelength is the same as that of the traveling waves.

All of these observations can be explained by remembering some trigonometry. We will start by writing the equations for the two incoming waves. Since the waves have the same amplitude and wavelength and differ only in the direction of travel, we see that these functions differ only in the sign of the term in the argument that involves the time. Therefore, the total disturbance is represented as a sum of two cosine functions

$$y(x,t) = y_1(x,t) + y_2(x,t) = A\cos[k(x-vt)] + A\cos[k(x+vt)]$$

You may recall that there is a trigonometric identity that relates the sum of two cosines to the product of two cosines.

$$\cos(a) + \cos(b) = 2\cos\left(\frac{a+b}{2}\right)\cos\left(\frac{a-b}{2}\right)$$

We will use this identity to rewrite the equation of the wave as a product of two cosines, one a function of space and one a function of time.

$$y(x,t) = A\cos(kx - \omega t) + A\cos(x + \omega t) = 2A\cos(kx)\cos(\omega t)$$

This mathematical expression describes exactly what we are seeing in Figure 24.5.

We have a stationary wave with an amplitude equal to the sum of the amplitudes of the incoming waves. This pattern is then multiplied by a factor that oscillates in time between minus one and plus one. In the next section we will discuss how to create such a standing wave.

24.6 Standing Waves: Part II

Figure 24.6 shows a wave being excited at one end of a string. The wave then travels down the string until it reaches the opposite end which is fixed in place. What will happen when the wave reaches the fixed end of the string? The answer is that it will be reflected back with the same amplitude and frequency. We now have two waves in the string,

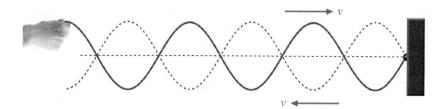

FIGURE 24.6 A wave is excited at the left end of the string and travels to the right until it is reflected from the fixed end of the string and will be reflected back (the dashed line). The resulting pattern in the string will be the sum of these two waves, yielding a standing wave.

moving in opposite directions with equal amplitudes and frequencies. These waves will then produce a standing wave pattern.

How can we understand the creation of this reflected wave? Recall that in the last unit we made the case that energy is transmitted by a traveling wave. What happens when that energy reaches the fixed end of the string? Since the end cannot move, it cannot absorb any energy. Consequently, the energy must remain in the string in the form of a reflected wave, much in the same way that a ball bounces off a fixed wall.

We can take this idea one step further and fix the initial end of the string as well (Figure 24.7). You will probably recognize this construction as that of any stringed musical instrument. In this case the fixed two ends of the string define the boundary conditions for all waves that can travel on the string. Since the displacement of the string has to be zero at each end, and we know that the separation between successive nodes of the wave is equal to half a wavelength, we can see that the distance between the fixed ends of the string must be an integral number of half wavelengths.

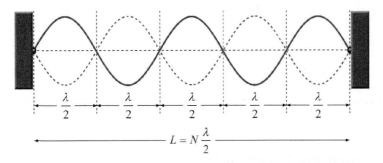

FIGURE 24.7 The standing waves that can be set up in a string with both ends fixed are determined by requiring that there be an integral number of half wavelenths in the total length of the string.

The tricky thing here is how to get the wave started since we are no longer hanging on to one end. Different musical instruments deal with this task in different ways. In a guitar, you pluck the string with your finger. In a piano, the string is struck with a small hammer. In a violin, the string is rubbed with a bow. Depending on how the string is manipulated, the standing wave that results will be a combination of many of the wavelengths that are possible. Usually, the biggest contribution comes from the longest possible wavelength the string can support. In this fundamental mode, the wavelength is twice the length of the string. Since the wavelength is fixed by the length of the string, and the speed of the wave is fixed by the properties of the string, such as its weight and its tension, the frequency of vibration of the string is also fixed. The different notes that can be played on the same string of a guitar or a violin are made by pushing the string against the neck of the instrument, effectively changing the distance between the fixed ends of the string. Changing this distance then changes the wavelength of the fundamental, which in turn changes the frequency that you hear.

24.7   Example: Tuning Your Guitar

The high E string on a typical guitar is usually made from 10 gauge wire, which has a mass per unit length of about a third of a gram per meter. Suppose the distance between the bridge and the nut of your guitar, in other words the length of string between the fixed end points, is 61 cm. We'd like to determine the tension in this string when the frequency of oscillation is 329.60 Hz.

How do we go about making this calculation? We will start by using the wavelength and frequency of the fundamental mode to determine the speed of the wave in the string. We can then use the mass of the string to calculate the value of the tension which gives the required speed.

We start by finding the wavelength of the fundamental mode, which is just twice the length of the string.

$$\lambda = 2L = 1.22 \text{ m}$$

The speed of the wave is always equal to the product of the wavelength and the frequency.

$$v = \lambda f = 402.1 \text{ m/s}$$

We know the speed of the wave is equal to the square root of the tension divided by the mass density of the string. Solving this equation, we obtain the required tension to be equal to 53.3 N.

$$v = \sqrt{\frac{T}{\mu}} \qquad \Rightarrow \qquad T = \mu v^2 = 53.3 \text{ N}$$

The frequencies of the six strings of a guitar vary from high to low as you move across the neck from bottom to top. To make the guitar easier to play and to keep the neck of the guitar from twisting, the tension in all six of these strings has to be about the same. Since all of the strings are the same length, the only way we can make the other strings vibrate with different frequencies is to have the speed of the waves on these strings be different. The strings that vibrate with a higher frequency must have a correspondingly higher speed. Since the tensions in the strings are about the same, the mass per unit length of the lower frequency strings must be bigger. This observation explains why the strings on a guitar get thicker and heavier as you move across the neck from higher to lower frequencies.

MAIN POINTS

Wave Equation Solutions

Any function whose argument is proportional to $(x \pm vt)$ is a solution to the wave equation with velocity equal to v.

$$y(x,t) = f(x - vt)$$ Waves traveling in the $+x$ direction

$$y(x,t) = f(x + vt)$$ Waves traveling in the $-x$ direction

The velocity of mechanical waves is obtained by applying Newton's second law to produce an appropriate equation of motion.

The velocity of the wave is determined totally by the properties of the medium.

Superposition and Standing Waves

$$y(x,t) = 2A \cos(kx) \cos(\omega t)$$

If more than one wave is excited in a medium, the resulting motion is given by the displacements of the individual waves at each point at each time.

If two waves of equal amplitude and frequency travel in the same medium, but with opposite velocities, the resultant motion will be a standing wave.

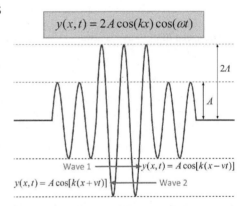

Wave 1 ⟶ $y(x,t) = A \cos[k(x - vt)]$

$y(x,t) = A \cos[k(x + vt)]$ ⟵ Wave 2

Standing Waves in a String

An initial wave propagated on a string that has both ends fixed will be reflected from the downstream fixed end, producing an additional wave which, when added to the initial wave, gives a standing wave.

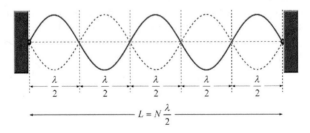

PROBLEMS

1. Guitar String: A guitar string with mass density $\mu = 0.00024$ kg/m is $L = 1.09$ m long on the guitar. The string is tuned by adjusting the tension to $T = 124.2$ N. (a) With what speed do waves on the string travel? (b) What is the fundamental frequency for this string? (c) Someone places a finger a distance 0.172 m from the top end of the guitar. What is the fundamental frequency in this case? (d) To "down tune" the guitar (so everything plays at a lower frequency) how should the tension be adjusted: *increase the tension, decrease the tension,* or *changing the tension will only alter the velocity not the frequency?*

2. Standing Wave: A standing wave pattern is created on a string with mass density $\mu = 0.00035$ kg/m. A wave generator with frequency $f = 61$ Hz is attached to one end of the string and the other end goes over a pulley and is connected to a mass (ignore the weight of the string between the pulley and mass). The distance between the generator and pulley is $L = 0.72$ m. Initially the third harmonic wave pattern is formed. (a) What is the wavelength of the wave? (b) What is the speed of the wave? (c) What is the tension in the string? (d) What is the mass hanging on the end of the string? (e) The hanging mass is adjusted to create the second harmonic. The frequency is held fixed at $f = 61$ Hz. What is the wavelength of the wave? (f) What is the speed of the wave? (g) What is the tension in the string? (h) What is the mass hanging on the end of the string? (i) Keeping the frequency fixed at $f = 61$ Hz, what is the maximum mass that can be used to still create a coherent standing wave pattern?

3. Wave on a Rope (INTERACTIVE EXAMPLE): An apparatus similar to the one used in lab uses an oscillating motor at one end to vibrate a long rope with frequency $f = 40$ Hz and amplitude $A = 0.25$ m. The rope is held at constant tension by hanging a mass on the other end. The rope has mass denstiy $\mu = 0.02$ kg/m and tension $T = 20.48$ N. Assume that at $t = 0$ the end of the rope at $x = 0$ has zero y displacement. What is the y displacement of the piece of rope at $x_1 = 0.5$ m when $t = 0$?

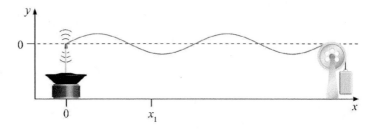

FIGURE 24.8 Problem 3

UNIT

25

FLUID STATICS

25.1 Overview

In this unit, we will begin our study of fluids. We will start with a discussion of the origins of pressure in a contained fluid. We will use Newton's second law to determine the relationship between this pressure and the depth below the surface of the fluid. We will then use this relationship to obtain Archimedes' principle: the relation between the buoyant force on an object and the volume it displaces when placed in a fluid. Finally, we will examine the conditions for sinking and floating, including a determination of the submerged fraction of a floating object.

25.2 States of Matter

So far in this course we have focused on systems involving solid objects (boxes, cylinders, springs and ramps, for example). We know that ordinary matter can also exist as liquids and gases, and we will devote the last two units of this course to understanding the similarities and the differences between the three states of matter in the context of the central ideas in this course, forces and motion.

Our understanding of the atoms that make up a solid object is that they are more or less rigidly fixed in position relative to each other. Applying a force at one point on a solid object will cause the object to move as a whole according to Newton's laws. The number and type of atoms contained in a solid object determine its mass. The shape of the object stays the same unless it is somehow bent or broken.

In a liquid or a gas, the positions of the molecules are *not* fixed; they can move relative to one another and assume a shape determined by their immediate environment. The molecules in a liquid bind weakly with each other in such a way that they are about as close together as they can get even though their positions are not fixed. The molecules in a gas do not bind with each other at all and tend to be much farther apart. The shape a liquid or gas takes is determined by their immediate environment. Despite these differences, many of the concepts that we will be developing in these two units will be identical for liquids and for gases. Indeed, the fact that neither of these is rigid means that we will need to re-examine the way we describe both forces and motion in problems involving these states of matter. We will begin this discussion by introducing the concept of pressure in the next section.

25.3 Pressure

Figure 25.1(a) shows a glass of water sitting on a table. The weight of the water causes the liquid to push outward on the inner surface of the glass, and the glass in turn keeps the water from spilling by pushing back inward on the liquid. These forces balance and the system is in static equilibrium. Even though the overall situation seems pretty simple, we need a better way to describe the forces involved before we can quantify them.

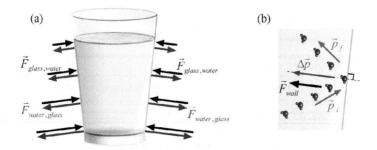

FIGURE 25.1 (a) The water exerts an outward force on the glass and the glass exerts an inward force on the water. These forces cancel, leaving the glass of water in a state of equilibrium. (b) A section of the water-glass interface is shown in which the origin of the force the glass exerts on the water is identified as the change in momentum that the molecules of water experience over the short time of their collisions with the glass.

The origin of the macroscopic force between the liquid and the wall is the collisions between the individual molecules of the liquid and the wall of the container as illustrated in Figure 25.1(b). Each time a molecule collides with the wall, its momentum changes in a direction perpendicular to the wall. Therefore, during the collision, there must have been a perpendicular force exerted by the wall on the molecule to cause its change in momentum.

Since we are interested only in macroscopic properties, we will account for these collisions by determining the average force acting over a time interval. We will then define the **pressure** as this average force per unit area on the wall

$$P \equiv \frac{F_{average}}{Area}$$

For example, if the average force of the water in a pool pushing against an underwater window is 10,000 N, and the area of the window is 2 square meters, then the pressure of the water on the window is 5,000 N/m². The same pressure acting on a window having twice the area would exert a total force on the window that was twice as big. The pressure of a liquid always produces a force that is perpendicular to the surface it touches.

Molecules making up a liquid will collide with any object, even each other, in exactly the same way that they collide with the walls of their container. Therefore, pressure is a concept that has meaning *everywhere* in the liquid, not just where it touches the container. For example, you can feel the pressure in your ears increasing as you dive to the bottom of a swimming pool. Indeed, we will discuss this change in pressure versus depth in a liquid in the next section.

25.4 Pressure and Depth

Figure 25.2(a) shows a tank containing some liquid. Our task is to determine how the pressure in the liquid varies with the depth beneath the surface.

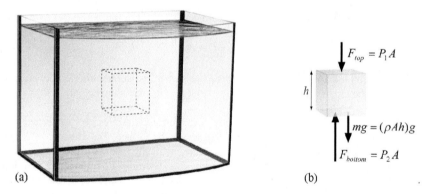

FIGURE 25.2 (a) A tank of water. (b) The vertical forces that act on a volume of the water.

Since the liquid in the tank isn't going anywhere on average, the net force on it must be zero. For the same reason, the net force on any part of the liquid we choose to focus on must also be zero. Figure 25.2(b) shows a small volume of the liquid inside the tank. The volume is a cube whose sides have height h.

Since the liquid is identical inside and outside the volume, the net force on the volume must be zero. What forces act on this volume? The weight of the volume, which can be determined from the density of the liquid and the volume of the cube, points down. The pressure of the liquid will exert a force on all six sides of the cube, but since these forces are always perpendicular to the sides, only the pressure acting on top and bottom surfaces will result in vertical forces. We can now use the fact that the net vertical force on the cube must be zero to find a simple relationship between the pressure difference between the top and bottom of the cube. Namely, the difference in pressure is simply proportional to the difference in depth.

$$P_2 A - P_1 A - (\rho A h)g = 0 \qquad \Rightarrow \qquad P_2 - P_1 = \rho g h$$

This result has two important features. First, it demonstrates that any two points at the same depth have to have the same pressure. Second, it defines the quantitative dependence of the pressure in the liquid on the depth. We will use both of these features in the next section when we work through the example of measuring atmospheric pressure.

25.5   Example: The Barometer

Figure 25.3(a) shows a long transparent tube, submerged in a swimming pool. Its top is sealed so that there is no air inside it. The sealed end of the tube is now raised out of the pool and the water inside the tube rises with it, as shown in Figure 25.3(b), just like it does if you cover the end of a straw as you pull it out of a glass of water. As the tube is pulled upward, a height is reached after which the level of the water in the tube stays constant as the tube is raised, as shown in Figure 25.3(c). Why does this happen?

We have just learned how pressure varies with depth in a given liquid.

$$P_2 - P_1 = \rho g h$$

We can use this knowledge to compare the pressure at the surface of the pool with the pressure at the top of the column of water when it is at its maximum height.

Since the surface of the water in the pool is stationary, the pressure just beneath the surface must be the same as the pressure just above the surface. The pressure, P_{atm}, just above the surface is the atmospheric pressure, typically about 10^5 N/m^2, or about 15 lb/in^2. We know that the pressure in the liquid is determined by the depth; therefore, the pressure P_2 at a point inside the tube, which is at the same height as the surface of the water just outside the tube, must also be equal to the atmospheric pressure.

Since there was no air at the top of the tube to begin with, the space that opens up above the water as the tube is raised contains no air at all, and the pressure there, P_1, must just be the vapor pressure of the water which is so much smaller than the outside atmospheric pressure that we can call it zero. Therefore, the difference in pressure at the water surface

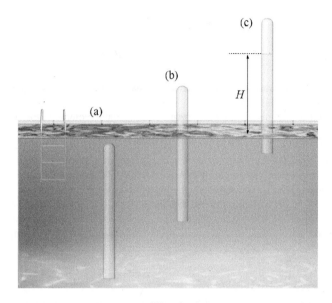

FIGURE 25.3 (a) A transparent tube whose top is sealed so that there is no air inside is submerged in a swimming pool. (b) The tube is pulled up and water inside rises with it. (c) At a certain height H, the level of the water ceases to rise.

and at the maximum height of the water in the tube must just be equal to the atmospheric pressure.

$$P_2 - P_1 = P_{atm} - P_1 \approx P_{atm} = \rho g H$$

This height then can be used to actually measure the atmospheric pressure. Such a device is called a **barometer**. Most barometers of this design use mercury rather than water since the density of mercury is much larger (almost 14 times larger) than that of water so that the heights involved are smaller. For example, typical atmospheric pressure (15 lb/in^2) corresponds to a height of about 29 inches of mercury, a phrase you've likely heard in weather forecasts.

25.6 Archimedes' Principle

Buoyancy is the upward force that an object experiences when it is placed in a liquid. This force determines whether an object floats or sinks when put in a liquid.

The buoyant force can be understood in terms of the ideas we used to derive the depth dependence of the pressure in a liquid. Namely, since the total force on any particular volume of the liquid must be zero, we determined that the net upward force due to the pressure of the liquid surrounding it, that is, the buoyant force, is just equal to the weight of the liquid in the volume, as shown in Figure 25.2(b). We say the buoyant force on any object placed in a liquid is just equal to the weight of the liquid displaced by that object, which is known as **Archimedes' principle**.

$$F_{buoyant} = F_{bottom} - F_{top} = (P_2 - P_1)A = g\rho_{liquid}V_{displaced} = \text{Weight of liquid displaced}$$

We now replace this volume of liquid with an equal volume of another material that has a greater density than the liquid, say steel. What will be the buoyant force on this object? It must just be the *same* as the buoyant force on the previous volume of the liquid since the force is just determined by the surrounding liquid which has not changed!

In this case, since the density of the material is greater than that of the liquid, the buoyant force will be less than the weight of the object and it will sink to the bottom of the tank! The force needed to lift the object off the bottom of the tank, however, is now less than the weight of the object, since the buoyant force is always present and always points up. In other words, if we suspend the object from a string while it is submerged in the liquid, we will again have a case of static equilibrium and the tension in the string will be *less than the weight of the box* by an amount exactly equal to the buoyant force.

25.7   Floating

We have just introduced the buoyant force and examined an example of this force exerted on an object whose density was greater than that of the liquid. What happens if we replace that object by another one whose density is less than that of the liquid, say a plastic box?

We start by completely submerging such a box in the liquid as shown in Figure 25.4. The weight of the box is now less than the weight of the displaced liquid, hence the net force on the box points upward causing it to rise. Once it reaches the surface it will simply float.

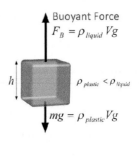

Buoyant Force
$$F_B = \rho_{liquid} Vg$$

h $\rho_{plastic} < \rho_{liquid}$

$$mg = \rho_{plastic} Vg$$

FIGURE 25.4 A plastic box is submerged in a liquid. The density of the plastic is less than the density of the liquid; therefore, the buoyant force exerted by the liquid on the box is greater than the weight of the box. When released, the box will rise.

Since the box is motionless once it floats on the surface, it must be in static equilibrium; therefore, the net force on the box must be zero. In particular, the upward buoyant force exerted on the box by the liquid must be exactly equal to the downward weight of the box as shown in Figure 25.5. We already know that the buoyant force is equal to the weight of the displaced liquid, therefore, an object that floats must be displacing exactly its own weight of the liquid!

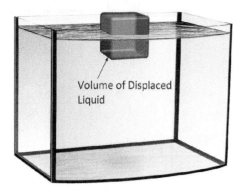

Volume of Displaced Liquid

FIGURE 25.5 The equilibrium position of the box from Figure 25.4. The volume of water displaced by the box is determined from setting the buoyant force equal to the weight of the box.

Since the density of the liquid in this case is greater than the density of the box, only part of the box will be under the surface of the water. To determine exactly how much of the box will be below the surface of the water, we just need to set the entire weight of the box equal to the weight of the liquid displaced, which is equal to the product of the density of the liquid and the submerged volume of the box.

$$\rho_{liquid} V_{displaced}\, g = \rho_{plastic} V_{plastic}\, g$$

$$V_{displaced} = \frac{\rho_{plastic}}{\rho_{liquid}} V_{plastic}$$

Thus, we see that the submerged fraction of the box just depends on the relative densities of the box and the liquid. The closer the density of the box is to the density of liquid, the bigger the submerged fraction of the box. For example, a piece of Styrofoam, which is much less dense than water, will float with very little of its volume submerged. On the other hand, an iceberg, which has a density that is about 90% that of water, will have about 90% of it volume submerged as it floats.

MAIN POINTS

Pressure and Depth Relationship

The pressure in a fluid is defined to be equal to the average force per unit area exerted by the molecules of the fluid during their collisions with the container walls.

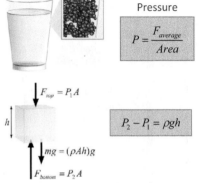

Pressure

$$P = \frac{F_{average}}{Area}$$

Setting the net force on any volume of the fluid, we obtain the result that the difference in pressure between two points in the fluid is proportional to the vertical separation of the two points.

$F_{top} = P_1 A$

h

$mg = (\rho A h)g$

$F_{bottom} = P_2 A$

$$P_2 - P_1 = \rho g h$$

Archimedes' Principle

The buoyant force exerted by the fluid on any object placed in it is determined from the relationship between the pressure and depth.

Archimedes' principle is derived from this determination of the buoyant force. Namely, the buoyant force exerted on an object is equal to the weight of the fluid it displaces.

F_B

mg

Buoyant Force

$$F_B = \rho_{liquid} V_{displaced} g$$

Floating

Archimedes' principle can be used to determine that the submerged fraction of a floating object is equal to the ratio of the mass density of the object to the mass density of the fluid.

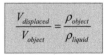

Submerged Fraction of a Floating Object

$$\frac{V_{displaced}}{V_{object}} = \frac{\rho_{object}}{\rho_{liquid}}$$

PROBLEMS

1. U Tube: A "U" shaped tube with a constant radius is filled with water and oil as shown below. The water is a height $h_1 = 0.37$ m above the bottom of the tube on the left side of the tube and a height $h_2 = 0.12$ m above the bottom of the tube on the right side of the tube. The oil is a height $h_3 = 0.30$ m above the water. Around the tube the atmospheric pressure is $P_{atm} = 101,300$ Pa. Water has a density of 1,000 kg/m^3. (a) What is the absolute pressure in the water at the bottom of the tube? (b) What is the absolute pressure in the water right at the oil-water interface? (c) What is the density of the oil? (d) Now the oil is replaced with a height $h_3 = 0.30$ m of glycerin which has a density of 1,261 kg/m^3. Assume the glycerin does not mix with the water and the total volume of water is the same as before. What is the absolute pressure in the water at the interface between the water and glycerin? (e) How much higher is the top of the water in the tube compared to the glycerin? (Labeled d in the diagram.) (f) What is the absolute pressure in the water at the very bottom of the tube?

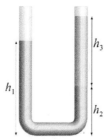

FIGURE 25.6 Problem 1 parts (a)-(c)

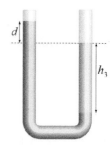

FIGURE 25.7 Problem 1 parts (d)-(f)

2. Balloon: A helium balloon ride lifts up passengers in a basket. Assume the density of air is 1.28 kg/m^3 and the density of helium in the balloon is 0.18 kg/m^3. The radius of the balloon (when filled) is $R = 4.5$ m. The total mass of the empty balloon and basket is $m_b = 125$ kg and the total volume is $V_b = 0.055$ m^3. Assume the average person that gets into the balloon has a mass $m_p = 73$ kg and volume $V_p = 0.078$ m^3. (a) What is the volume of helium in the balloon when fully inflated? (b) What is the magnitude of the force of gravity on the entire system when there are no people? Include the mass of the balloon, basket, and helium. (c) What is the magnitude of the buoyant force on the entire system when there are no people? Include the volume of the balloon, basket, and helium. (d) What is the magnitude of the force of gravity on each person? (e) What is the magnitude of the buoyant force on each person? (f) How many people can the balloon lift? Your answer must be an integer.

3. Floating Cylinders (INTERACTIVE EXAMPLES): Two cylinders with the same mass density $\rho_C = 713$ kg/m^3 are floating in a container of water (with mass density $\rho_W = 1,025$ kg/m^3). Cylinder 1 has a length of $L_1 = 20$ cm and radius $r_1 = 5$ cm. Cylinder 2 has a length of $L_2 = 10$ cm and radius $r_2 = 10$ cm. If h_1 and h_2 are the heights that these cylinders stick out above the water, what is the ratio of the height of cylinder 2 above the water to the height of cylinder 1 above the water (h_2 / h_1)?

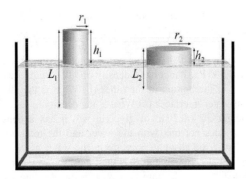

FIGURE 25.8 Problem 3

UNIT

26

FLUID DYNAMICS

26.1 Overview

In this unit, we will conclude our study of fluids by extending our study to systems in which the fluid is flowing. We will limit our discussion to ideal fluids, namely the smooth flow of incompressible, zero-viscosity fluids. We will first develop the continuity equation that relates the speed of the flow to the cross-sectional area of the pipe. We will then move on to consider flow in pipes whose area and height can change to derive Bernoulli's equation that relates the pressure change in such flow to the change in the kinetic and gravitational potential energy density of the fluid.

26.2 Moving Fluids

To this point our discussion of liquids and gases has been limited to systems in *static equilibrium*, allowing us to understand how pressure changes with depth and to quantify the buoyant force on objects which are partially or totally submerged.

In this unit we will use Newton's laws to extend our treatment to systems in which the liquid or the gas is flowing. Understanding the flow of fluids is, in general, a very complicated problem unless some simplifying assumptions are made. While these assumptions will limit the range of actual situations to which we can apply our equations, the development and the form of these equations will still give us great insight into the behavior of liquids in general.

In what follows, we will assume that we are dealing with **ideal fluids**. An ideal fluid is a liquid or a gas with the following properties:

1) It is incompressible. By incompressible we mean that the density of the fluid cannot change. This assumption is very good for many liquids, like water for example, but it can even be a reasonable approximation for gases in certain situations.

2) It has zero viscosity, that is, it flows *without* friction. It flows smoothly and without turbulence.

26.3 Continuity Equation

One of the defining assumptions for an ideal fluid is that it is incompressible. This simple property alone will yield a very important relationship describing the properties of moving fluids, called the continuity equation. If a fluid is incompressible, then a certain mass of the liquid *always* occupies the same volume, no matter the shape of its container.

Figure 26.1 shows water flowing through a pipe consisting of two cylinders having different cross-sectional areas joined by a transition piece. Since water is impressible, the amount of water flowing past *any* part of the pipe during a specific time interval must be the same!

FIGURE 26.1 Water flows in a pipe consisting of two cylinders having different cross-sectional areas joined by a transition piece.

This result is a simple consequence of the fact that whatever goes into one end must come out the other! Since the fluid is incompressible, the density of the fluid must be the same anywhere in the pipe. The amount of water dm that flows past a point in the pipe in a time dt is just equal to the volume of the column of water that passes in this time, which is equal to the product of its area and its length (see Figure 26.2).

$$dm = \rho\, dV_i = \rho A_i\, dl_i \qquad \Rightarrow \qquad dV_1 = dV_2 \qquad \Rightarrow \qquad A_1\, dl_1 = A_2\, dl_2$$

FIGURE 26.2 The amount of water dm that flows past any point in the pipe in a time dt is equal to the volume of the column of water that passes in this time, which is equal to the product of its area (A_i) and its length (dl_i).

Since the length of the volume is equal to the product of the speed of the flow and the time interval ($dl_i = v_i\, dt$), we obtain the important result that the product of the area of the pipe and the speed of the flow must be the same for all parts of the pipe.

$$A_1 v_1 = A_2 v_2$$

This equation is called the **continuity equation**.

Thus, we see that as the diameter of a pipe decreases, the speed of the flow through it must increase. You are probably already familiar with this result if you have ever put a small nozzle at the end of a garden hose in order to increase the speed of the water leaving the hose.

We will now do a simple example to illustrate the usefulness of the continuity equation. Figure 26.3 shows a pipe in the basement of your house that carries water in from the city supply. As you open up a faucet to your bathtub, you see the water flowing out of the pipe at 4 m/s. If the diameter of the upstairs pipe is half that of the basement pipe, what is the speed of the water in the basement pipe?

In order to answer this equation, we simply apply the continuity equation. Namely, the product of the area of the basement pipe and the speed of the flow there must be equal to the product of the area of the faucet and the speed of the flow at the bathtub.

$$A_1 v_1 = A_2 v_2 \qquad \Rightarrow \qquad v_1 = v_2 \frac{A_2}{A_1}$$

Since the area of the pipe is proportional to the square of its diameter, we see that the speed of the flow in the basement must be one-fourth of the speed of the flow at the bathtub, or 1 m/s.

$$v_1 = v_2 \frac{\pi d_2^2}{\pi d_1^2} = v_2 \frac{d_1^2/4}{d_1^2} = \frac{1}{4} v_2 = 1 \text{ m/s}$$

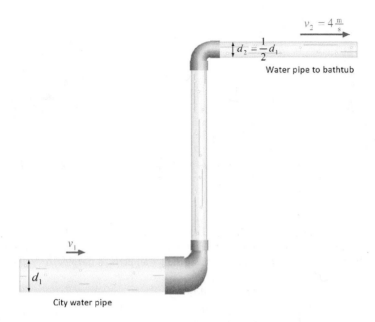

FIGURE 26.3 Water flows from the city water pipe to an upstairs bathtub. The continuity equation can be used to determine the ratio of the flow velocities in the two pipes in terms of the diameters of the pipes.

26.4 Energy Conservation

We have just shown that the speed of the fluid in a pipe needs to change whenever the diameter of the pipe changes in order to keep the flow rate the same. If the speed of the water is changing then so is its kinetic energy. If its kinetic energy is changing, then there must be forces acting that do work on the water. What are these forces? How can we quantify them?

We will start by looking at the flow in a pipe whose cross-sectional area is reduced from A_1 to A_2 somewhere along the flow as shown in Figure 26.1. We have colored a volume of the water red in Figure 26.4.

The transition shown from Figure 26.4(a) to Figure 26.4(b) corresponds to the flow of the water during a time dt. In particular, during the time it takes a certain volume of water to flow past point 1, the same volume of water will flow past point 2. The change in kinetic energy of the red water during this time will be the kinetic energy of the water that flows past point 2 minus the kinetic energy of the water that flows past point 1.

$$\Delta K = \frac{1}{2}m_2 v_2^2 - \frac{1}{2}m_1 v_1^2$$

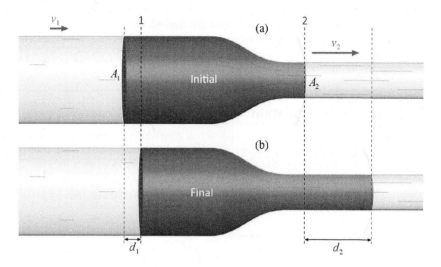

FIGURE 26.4 (a) The initial state: A portion of the water that is colored red extends through both regions of the pipe. (b) The final state: A time dt later, the colored volume has moved to the right, vacating a length d_1 in the large diameter portion and filling a length d_2 in the smaller diameter portion of the pipe.

The work done on the red water by the pressure in the left part of the pipe is equal to the product of the force exerted by this pressure and the distance the water moves.

$$W_1 = F_1 d_1 = (P_1 A_1) d_1$$

In the same way, the work done on the red water by the pressure in the right part of the pipe is equal to the product of the force exerted by this pressure and the distance the water moves.

$$W_2 = -F_2 d_2 = -(P_2 A_2) d_2$$

Note that the work done by the pressure on the right is negative since the force due to this pressure points in the opposite direction of the flow of the water. We will now use the work-kinetic energy theorem to equate the forms for the change in kinetic energy and the work done by the pressure:

$$\frac{1}{2} m_2 v_2^2 - \frac{1}{2} m_1 v_1^2 = P_1 A_1 d_1 - P_2 A_2 d_2$$

This equation can be simplified by noting that both the masses and the volumes involved at points 1 and 2 are equal!

$$\frac{1}{2} \rho V (v_2^2 - v_1^2) = (P_1 - P_2) V \qquad \Rightarrow \qquad P_2 + \frac{1}{2} \rho v_2^2 = P_1 + \frac{1}{2} \rho v_1^2$$

Note that our final result implies that at points in the flow where the pressure is larger, the velocity must be smaller. We can understand this dependence in terms of the forces that are acting; namely, as the fluid moves from a region of high pressure to one of low pressure it speeds up since the direction of the net force is in the same direction as the movement of the liquid, causing it to accelerate.

26.5 Bernoulli's Equation

So far we have considered flow along a horizontal pipe where the only work done on the fluid was due to differences in pressure. We can easily generalize this result to the case where the fluid flows in a pipe whose vertical height is changing, as shown in Figure 26.5, by just including the work done on the fluid by gravity. In this case the change in kinetic energy of the flowing liquid is equal to the sum of the work done by both pressure and gravity.

$$\Delta K = W_{pressure} + W_{gravity}$$

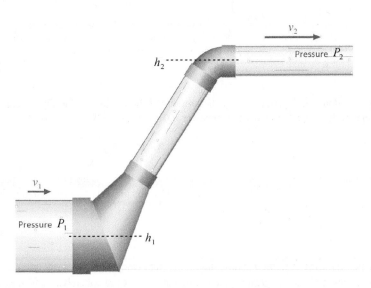

FIGURE 26.5 A fluid flows in a pipe with changing cross-sectional area and height. The relation between the pressure and the velocity of the flow at any point in the pipe can be determined from the work-kinetic energy theorem by including the work done on the fluid by gravity. The result is called Bernoulli's equation.

The derivation is very similar and results in the addition of a term on either side of the equation that represents the gravitational potential energy per unit volume of the fluid.

$$P_1 + \frac{1}{2}\rho v_1^2 + \rho g h_1 = P_2 + \frac{1}{2}\rho v_2^2 + \rho g h_2$$

This equation is often referred to as **Bernoulli's equation**. Note that, in the static limit (that is, $v_1 = v_2 = 0$), we recover our result from the last unit that relates the pressure to the depth! In general, we see that the increase in pressure is equal to the decrease in the sum of the kinetic and gravitational potential energies per unit volume.

26.6   Example: Bernoulli's Equation

We will close with a problem that illustrates the use of Bernoulli's equation. Figure 26.6 shows a pipe in the basement of your house that carries water in from the city supply. Up on the first floor, 4 meters above the basement pipe, you open up a faucet to your bathtub and see the water flowing out of the pipe at 4 m/s. The diameter of the upstairs pipe is half that of the basement pipe. What is the difference in water pressure between the basement and the upstairs?

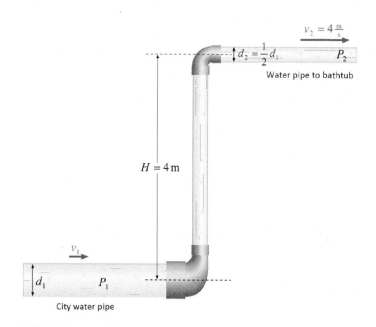

FIGURE 26.6 Bernoulli's equation can be used to determine the difference in pressure between the inlet city water pipe and the outlet pipe to the bathtub shown.

We start by writing down Bernoulli's equation:

$$P_1 + \frac{1}{2}\rho v_1^2 + \rho g h_1 = P_2 + \frac{1}{2}\rho v_2^2 + \rho g h_2$$

We can solve this equation directly for the pressure difference between the input and the bathtub.

$$P_1 - P_2 = \frac{1}{2}\rho\left(v_2^2 - v_1^2\right) + \rho g\left(h_2 - h_1\right)$$

The only term we cannot immediately evaluate on the right hand side of this equation involves the velocity of the incoming water. We can evaluate this term if we use the continuity equation that relates the ratio of the velocities to the ratio of the square of the diameters of the pipes.

$$A_1 v_1 = A_2 v_2 \qquad \Rightarrow \qquad \frac{v_2}{v_1} = \frac{d_1^2}{d_2^2}$$

Indeed, we found this velocity to be equal to 1m/s in an earlier example in this unit. If we now plug in the remaining numbers we find that the pressure in the basement pipe is bigger than the pressure upstairs by 46,700 N/m², or 46,700 pascals (Pa), the SI unit for pressure.

MAIN POINTS

Continuity Equation

We used the fact that an ideal fluid is incompressible (i.e., its density does not change) to derive the continuity equation: The product of the speed of the flow and the cross-sectional area of the pipe through which an ideal fluid flows is a constant!

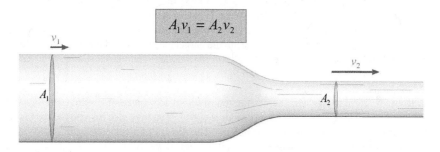

$$A_1 v_1 = A_2 v_2$$

Bernoulli's Equation

We used the work-kinetic energy theorem to determine that the sum of the work done by the pressure of the fluid and kinetic energy density is a constant at any point in the flow.

By considering ideal fluids that flow from one height to another, we used the work-kinetic energy theorem to derive Bernoulli's equation: that the change in pressure of an ideal fluid between two points is equal to the sum of the changes in the kinetic energy density and the gravitational potential energy density of the fluid.

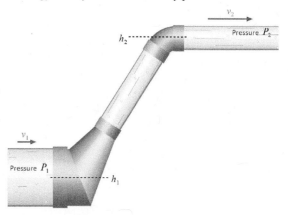

$$P_2 - P_1 = \frac{1}{2}\rho(v_1^2 - v_2^2) + \rho g(h_1 - h_2)$$

PROBLEMS

1. Pipe: Water flows in a pipe that tapers down from an initial radius of $R_1 = 0.21$ m to a final radius of $R_2 = 0.12$ m. The water flows at a velocity $v_1 = 0.8$ m/s in the larger section of the pipe. (a) What is the volume flow rate of the water? (b) What is the velocity of the water in the smaller section? (c) Using this water supply, how long would it take to fill a swimming pool with a volume of $V = 150$ m^3? Give your answer in minutes. (d) The water pressure in the center of the larger section of the pipe is $P_1 = 273{,}510$ Pa. Assume the density of water is 1,000 kg/m^3. What is the pressure in the center of the smaller section of the pipe? (e) If the pipe was turned vertical, such that the larger section were on the bottom and the volume flow rate in the larger section was kept the same, which answers would change: *the speed of water in the smaller section*, *the volume flow rate in the smaller section*, and/or *the pressure in the smaller section*?

2. Roof: A storm blows wind across the top of a roof at a speed of $v = 29.4$ m/s. Assume the air on the lower side of the roof (inside) is at rest. You may also assume the roof is flat (i.e., zero pitch) and has an area $A = 38.8$ m^2. The density of air is 1.28 kg/m^3. (a) What is the pressure differential across the top and bottom of the roof? (b) What is the net lift force on the roof due to the storm?

UNIT

27

INTRODUCTION TO THERMODYNAMICS

27.1 Overview

We are about to embark on a journey into the realm of **thermodynamics**. The word itself suggests that we will be learning about heat and changes in temperature. While this is certainly part of the story, we will see that a thorough study of thermodynamics gives us much more.

First and foremost, we will be able to fill in the gaps that we had to gloss over in our study of mechanics. The biggest gap was our lack of a general rule that determines the equilibrium state of a system. That is, even though, in many cases, we know very well how things end up (liquids at the bottom of bowls, blocks at rest on surfaces, etc.) we had absolutely no general rule to apply in all cases. Thermodynamics will now give us such a rule.

One of our first steps will be to extend our discussion of energy conservation beyond the sum of mechanical kinetic and potential energies, where the kinetic energy was defined by the motion of the center of mass of an object and the rotation of the object around its center of mass.

While this approach is perfectly correct it is also incomplete since it ignores the *internal energy* that the object has due to the random motion of its individual atoms. In our study of a block sliding to rest due to friction, for example, we saw that the total mechanical energy of the block was not conserved, and we were told that the kinetic energy of the center of mass of the block was translated into heat which raised the temperature of both the block and the floor.

At the time we had no way to quantify this statement further, but that is all about to change. What's more, in addition to tying up the loose ends in mechanics we will soon understand the relationships between temperature, internal energy, and heat; we will learn about the thermal properties of solids, liquids, and gases; we will understand how engines work and what limits their efficiency; and we will see how all of the above are related to the key thermodynamic concept of entropy.

27.2 Mechanics Is Incomplete

We'll start with a simple example. Figure 27.1(a) shows two clay balls attached at both ends of a string which is initially taut and at rest. A constant force $\vec{F}$ is applied at the midpoint of the string in a direction perpendicular to its initial orientation. Figure 27.1(b) shows the balls just prior to the collision. At this point the total kinetic energy of the balls has the two components we are familiar with from mechanics: the kinetic energy due to the motion of the center of mass itself, and the kinetic energy of the motion of the two balls relative to the center of mass. At any time before the collision the sum of these two kinetic energy components will be exactly equal to the work done on the system by the applied force.

$$W_{Net} = FD = K_{CM} + K_{REL}$$

The kinetic energy of the center of mass of the balls is just equal to the applied force times the distance moved by the center of mass; therefore, the kinetic energy of the balls relative to the center of mass at any instant before the collision is just equal to the product of the applied force and the difference between the two distances.

$$K_{REL} = F(D - d_{CM})$$

The inelastic collision of the clay balls does not change the motion of the center of mass since no external forces are involved, so the kinetic energy of the center of mass will always be given by the same simple expression $K_{CM} = F d_{CM}$. The kinetic energy of the balls relative to the center of mass becomes zero during the collision, however, so we are

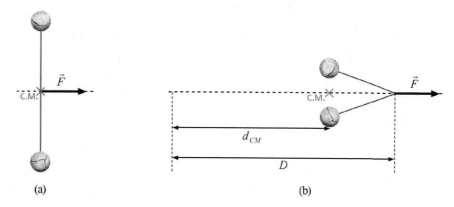

FIGURE 27.1 (a) A constant force $\vec{F}$ is applied at the midpoint of a string connecting two clay balls. (b) Just prior to the collision of the balls, the force $\vec{F}$ has acted over a distance D, which is larger than d_{CM}, the distance the center of mass has moved.

suddenly in a situation where the kinetic energy of the system is considerably *less than* the total work done on the system by the applied force. Where did this energy go?

The answer is that it became energy internal to the balls themselves. This internal energy is composed of the kinetic energies of the vibrating atoms making up the balls, plus the potential energies of the bonds holding these atoms together. The change of this internal energy is exactly equal to the difference between the total work done on the system and the change in the kinetic energy of its center of mass.

$$\Delta E_{Internal} = W_{Net \ on \ System} - \Delta K_{CM}$$

In other words, when internal energy is taken into account the total energy of any isolated system is *always* conserved. This is a simpler yet much more powerful statement than we can get from mechanics alone, and we will devote the rest of this course to understanding and quantifying it.

27.3   The First Law of Thermodynamics

As we showed in the last section, by taking into account the internal energy of a system we can arrive at a simple energy conservation relationship that should always be true, namely that the work done on a system by external forces will equal the change in total mechanical energy of the center of mass of the system plus the change in its internal energy.

$$W_{Net \ on \ System} = \Delta K_{CM} + \Delta E_{Internal}$$

We will do more with this equation later, but let's start with a simple example. Suppose we apply a constant horizontal force of 45 N to a 10 kg box, causing it to accelerate from rest across the floor. After the box has moved a distance of 2 meters, we remove the force and the friction between the floor and the box causes the box to slow down and eventually stop. How much does the internal energy of the box and the floor change during this process?

Since the applied force and the displacement of the box are parallel, the work done on the system by the external force is just equal to (45 N)(2 m) = 90 J. The change in total mechanical energy of the center of mass of the system is zero since the box starts and ends at rest and there is no change in its height. The change in internal energy of the system is therefore equal to 90 J.

This is a simple example, but it illustrates a very important subtlety, which is that we need to be very careful when we define what we mean by the "system". In this example, "the system" was made up of both the box and the floor, and the work that was done by the external force changed the internal energy of both. In other words, both the box and the floor were warmed up a bit.

In most thermodynamics problems, the total mechanical energy of the center of mass of the system will not change. For this reason it is customary to leave this part out of the equation altogether, making it appear even simpler.

$$W_{Net} = \Delta U$$

Note that we have used the symbol U here to represent the **internal energy** of the system, as opposed to its previous use as a potential energy

This is not quite the end of the story, however, since there is another very simple way that the internal energy of a system can be changed: We can just add energy to the system directly in the form of **heat**. Denoting the heat flowing into a system by the letter Q we can put everything together and arrive at the **first law of thermodynamics**:

$$\Delta U = W_{Net} + Q$$

This law expresses the fact that we can change the total internal energy of a system in two ways: We can do work on the system and we can add heat to the system. We will expand on this idea in the next few sections as we introduce the concept of temperature and discuss heat flow in more detail.

27.4   Irreversible Processes

As we have just seen, our study of thermodynamics has required us to generalize our concept of energy that we developed in the mechanics course so that we could introduce a full blown conservation of energy principle. Namely, energy is always conserved! The first law of thermodynamics tells us that the change in the internal energy of a system is equal to the work done on the system plus the heat that is added to the system. The internal energy is a property of the system; it can change by having energy (in the form of work and/or heat) flow into or out of the system. This is a powerful law that we will make great use of in this section.

However, from our own experience, we also know that this can't be the whole story. Consider the example of the block sliding to a stop described earlier. Once the external force was removed, the system (block plus floor) had only kinetic energy. When the block comes to rest, the system no longer has any obvious kinetic energy, but it does have an equivalent amount of internal energy!

So far so good, but why can't all of that internal energy be converted back into kinetic energy? What prevents the block and floor from cooling down, allowing the block to begin to move, ultimately regaining all of its initial kinetic energy as the internal energy of the block and floor return to its original value?

You know this process will never happen on its own, but it DOES NOT VIOLATE ENERGY CONSERVATION!

Let's look at another example. Suppose we place an ice cube on a sidewalk on a warm day. You know the fate of this ice cube. It will first melt, turning into water. This water will then evaporate into the atmosphere, leaving no trace of the original ice cube on the sidewalk. Energy was transferred from the environment to the ice cube to cause this transformation.

You know this process will also never happen in reverse. On a warm day, water vapor molecules in the atmosphere will never spontaneously move to the sidewalk forming a puddle of water which then solidifies into an ice cube!

This reverse process, however, does NOT violate energy conservation! We need something more to tell us how energy flows in this system. This "something more" will turn out to be a new universal principle, the second law of thermodynamics.

27.5   The Arrow of Time

In the last section we saw two examples of irreversible processes. That is, unlike most examples from our idealized treatment of mechanics, these processes indicate a direction for time. For example, if we make a movie of some elastic collisions of billiard balls and then run it backwards, it would be very difficult to tell which movie corresponds to the actual physical situation. That is, we say these processes are reversible. There is nothing in these processes that tells us the direction of time!

What is different for the irreversible processes discussed in the last section? Our natural inclination is to look at these processes microscopically to find the mechanisms responsible for the irreversibility. This task is daunting for two reasons. First, the number of particles involved is huge, typically larger than 10^{20}, and second, the individual processes are not simply described.

For example, if we could zoom in on the block as it comes to a stop, we would see the points of contact aren't moving along exactly with the center of mass of the block.

As these little parts of the surface are set vibrating, sound waves are sent into the block. These sound waves carry energy into the block which results in distortions of the bonds and an increase in the internal energy of the block. Rather than trying to describe the details of these microscopic interactions, we will account for the direction of this energy flow by introducing a new general principle that all processes must obey (the second law of thermodynamics) that is NOT derived from mechanics.

In order to present the second law of thermodynamics we will need to introduce a new quantity called entropy, which we will do in the next section. For now, we just want to present the big idea behind this new principle: that irreversibility is explained by having energy flow so that at equilibrium, the number of states available to the system is a maximum. For now, we'll just use a qualitative notion of the number of states as the number of ways the molecules can be rearranged. In our example, the molecules of ice are restricted to a certain volume, while the water vapor molecules can be located over a huge volume. The same number of water molecules have many more ways to arrange themselves in the gas than they do in the solid. Equilibrium then is reached by having the ice melt and the water evaporate. Surprisingly, the same principle- reaching the most numerous batch of states- governs how everything works, even, for example when ice condenses out from the atmosphere on a cold day. We will now formalize these ideas in the next section.

27.6   Equilibrium, Entropy, and the Second Law

We saw in the last section that we can explain the fate of an ice cube on a warm day by applying the general principle that equilibrium of a system corresponds to the situation that has the largest number of possible states of its microscopic components.

We will now formalize this new approach by introducing the concept of entropy and the second law of thermodynamics. Namely, we define the **entropy** of a system as being proportional to the logarithm of the number of possible states of its components. The **second law of thermodynamics** states that the entropy of an isolated system never decreases.

We will now illustrate these new ideas with a couple of simple examples. Figure 27.2(a) shows 100 distinct molecules bouncing around freely in a box, while Figure 27.2(b) shows the same molecules after a gate to an empty adjacent box of the same volume is opened. The gas molecules now spread out to fill both boxes.

You won't be surprised to learn that the number of states available to an individual molecule in both boxes is just twice the number available to it when it was confined to one box. The total number of states available to the system of 100 molecules in the first box is just equal to the product of the number of states available to each molecule. That is, if we denote the number of states available to any individual molecule in the first box as Ω, then the total number of states available to our system of 100 molecules is equal to Ω^{100}. When both boxes are open, the total number of states increases to $(2\Omega)^{100}$, a huge number, an increase of a factor of 2^{100} or about a factor of 10^{31}. Using the properties of logarithms, we see that entropy increases in this irreversible process by a factor of 100 ln(2), or about a factor of 70.

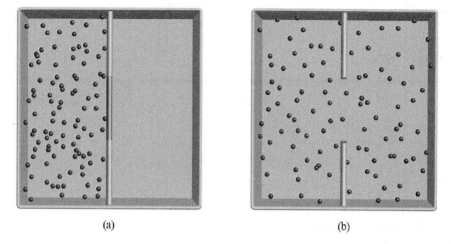

(a)   (b)

FIGURE 27.2 (a) 100 distinct molecules moving in a single box. (b) 100 ditinct molecules moving through both boxes after gate was opened.

27.7   Equilibrium from Maximizing Entropy

We saw in the last section that equilibrium was reached when the particle densities on either side of the piston were equal. The important feature of that exercise was not really the result, but rather the method we used to obtain the result. We did not use a model to describe the interactions of the molecules with the piston to make the calculation; we simply maximized the entropy of the system. The important take away message here is that this procedure always works!

For example, we will show in a later unit, how this procedure can be applied to two objects exchanging heat energy rather than volume to determine that heat will flow from the hot object to the cold object until they reach the same temperature. In fact, we can use this principle to actually define the thermodynamic temperature in terms of the derivative of the entropy of a system with respect to its internal energy.

In conclusion then, we see that we now have a general principle (the second law of thermodynamics) that nature always flows to maximize the total entropy. We account for irreversible processes by stating that the entropy of an isolated system never decreases. Entropy does not change during a reversible process and always increases during an irreversible process.

MAIN POINTS

Internal Energy

The equations that come from integrating Newton's second law accurately describe the motion of the center of mass of a system, but a complete description of the system must include accounts of the internal energy of the system.

First Law of Thermodynamics

The first law of thermodynamics is a statement of the conservation of energy that goes beyond that introduced in mechanics. In particular, the change in the internal energy of a system is equal to the work done on the system by external forces plus the energy, in the form of heat, that flows into the system.

First Law of
Thermodynamics

$$\Delta U = W_{Net} + Q$$

Second Law of Thermodynamics

The second law of thermodynamics is needed to account for irreversible processes. That is, there are processes that are allowed by energy conservation that never happen. There is a natural flow of events towards an equilibrium condition in which the number of possible states of the microscopic components of a system is maximized. We formalized this principle by defining the entropy of a system to be proportional to the logarithm of the number of its available microscopic states. The second law, then, states that nature always flows to maximize the total entropy. The entropy of an isolated system never decreases. Entropy does not change during a reversible process and always increases during an irreversible process.

Entropy

$$S \propto \ln \Omega$$

Second Law of
Thermodynamics

$$\Delta S \geq 0$$

28

HEAT AND TEMPERATURE

28.1 Overview

In the last unit, we presented the introduction to our approach to thermodynamics that will be presented in subsequent units. In this unit, we will be looking in more detail at the laws and concepts that were introduced in that introduction.

In particular, by applying the second law of thermodynamics to a system composed of some helium gas in an insulating container with the top sealed by a metal piston, we will be able to define an absolute temperature scale in terms of the fundamental concepts of entropy and energy. That is, by combining the fact that heat flows from the warmer gas to the cooler piston until they reach the same temperature with the fundamental concept that the entropy of the combined system reaches its maximum value at equilibrium, we can identify the temperature as the inverse of the derivative of the entropy with respect to the energy.

Finally, we will introduce the concepts of heat capacity and latent heats of vaporization and fusion that quantitatively determine how much energy is necessary to change either the temperature of a substance or its phase.

28.2   The First Law of Thermodynamics

In the last unit we introduced the first law of thermodynamics that states for systems in which the total mechanical energy of the center of mass of the system does not change, the change in the internal energy of the system is equal to the sum of the work done on the system by external forces and the heat that is added to the system.

$$\Delta U = W_{Net} + Q$$

Figure 28.1 shows a constant force being applied to a piston that compresses the helium gas in an insulating cylinder. This force does work on the on the gas equal to the integral of the force exerted by the piston on the gas over the distance the piston travels.

$$W_{Net} = \int_{y_i}^{y_f} \vec{F}_{Pistong, Gas} \cdot d\vec{y}$$

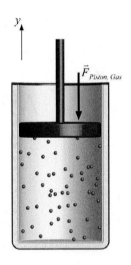

From Newton's third law, the force that the piston exerts on the gas is equal to minus the force that the gas exerts on the piston. The gas exerts a force on the piston that is determined by the pressure of the gas.

That is, the force the gas exerts on the piston is equal to the product of gas pressure with the cross-sectional area of the piston.

$$F_{Gas, Piston} = PA$$

Therefore, the work done by the piston on the gas can be written totally in terms of macroscopic properties of the gas, its pressure P and its volume V.

$$W_{Net} = -\int_{V_i}^{V_f} P \, dV$$

FIGURE 28.1 A constant force $\vec{F}$ is applied to a piston that compresses the helium gas in the insulating cylinder.

Note that if we fix the piston so that the volume of the gas does not change, no work can be done on the gas. Nonetheless, it is possible for the internal energy of the gas to change due to heat transfer between the gas and the piston via interactions of the molecules of the gas and the piston if the temperature of the piston is not the same as the temperature of the gas. We will soon show that this energy will flow from one system to the other until the temperatures of the gas and the piston are the same.

Now, we've been using the word temperature a lot because it is a common word that everyone knows, but we have not yet defined temperature in terms of the other thermodynamic concepts used in the two fundamental laws we introduced in the last unit. In the next two sections, we will develop such a definition.

28.3   Equilibrium and Temperature

We start from the stationary piston and gas system discussed in the last section. We will assume that the temperature of the gas is *higher* than that of the piston. We know what will happen. Energy will be transferred between the gas and the piston in the form of heat. Heat will flow from the hot gas to the cold piston. When will this heat flow stop? What will be the equilibrium condition?

As heat flows from the gas to the piston, the energy of the piston increases by the same amount that the energy of the gas decreases. The temperature of the piston increases as the temperature of the gas decreases. This process will cease when the temperatures of the gas and the piston become equal; this is the equilibrium condition.

Though this description certainly makes sense, it also is incomplete from a fundamental physics point of view. Namely, the condition for equilibrium involves temperature in a completely central way, but we have not yet defined temperature! You all have a good intuitive sense of temperature from your interactions with your environment, but how do we understand it in terms of the fundamental physical properties?

To answer this question, we will invoke the second law of thermodynamics that we introduced in the last unit. Namely, since the entropy of an isolated system never decreases, heat will flow between the gas and the piston until the total entropy of the two is maximized, at which point they will be in equilibrium! In the next section, we will apply this procedure to develop a definition for temperature.

28.4   Thermodynamic Temperature

In this section we will develop the definition of temperature. We start with the claim that the entropy of a system increases as we add energy to that system. We certainly don't know the exact dependences yet, but it is reasonable that entropy increases with energy since more states will become accessible at higher energies. Figure 28.2 shows the entropy as a function of energy for the piston and the gas.

Now, we know that all energy that leaves the piston will enter the gas such that the total energy of the gas-piston system will remain constant. Therefore, we can relate the energy of the gas to that of the piston and this constant total energy of the gas plus piston.

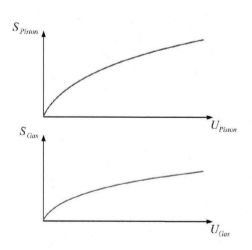

FIGURE 28.2 Plots of entropy vs. energy for the stationary piston-gas system.

$$U_{Gas} = U_{Total} - U_{Piston}$$

Consequently, we can represent all of the entropies on a single plot using the energy of the piston as the independent variable. Figure 28.3 shows this plot. We see that this total entropy plot does indeed have a maximum that defines the equilibrium point just as we expected from the second law of thermodynamics.

We now want to look at these plots in more detail to understand what drives this energy flow. In Figure 28.4 we have marked the initial values on the entropy vs. energy plots. We note that the slope (i.e., dS / dU) of the piston curve is *larger* than the slope of the gas curve. We claim that heat will flow from the system with the small slope (the gas) to the system with the large slope (the piston), so that the total entropy of the system will increase!

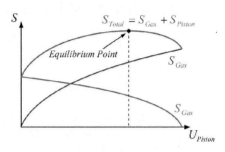

FIGURE 28.3 Plots of entropies vs. energy of piston in the stationary piston-gas system.

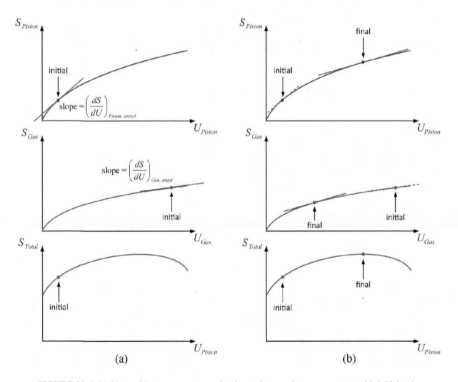

FIGURE 28.4 (a) Plots of entropy vs. energy for the stationary piston-gas system with initial values indicated. (b) Plots of entropy vs. energy for the system with initial and final values indicated. Note that the slopes (dS / dU) of the gas and piston plots are identical at equilibrium.

To see how this works, we write down the expression for change in entropy for the stationary piston-gas system.

$$\Delta S_{Total} = \left[\left(\frac{dS}{dU} \right)_{Gas} \times \Delta U_{Gas} \right] + \left[\left(\frac{dS}{dU} \right)_{Piston} \times \Delta U_{Piston} \right]$$

We now use the fact that $\Delta U_{Gas} = -\Delta U_{Piston}$ to obtain:

$$\Delta S_{Total} = \Delta U_{Piston} \times \left[\left(\frac{dS}{dU} \right)_{Piston} - \left(\frac{dS}{dU} \right)_{Gas} \right]$$

Since we know ΔU_{Piston} must be positive (initially, the gas is hotter than the piston), it must be true that $(dS/dU)_{Piston}$ is greater than $(dS/dU)_{Gas}$ in order for ΔS_{Total} to increase!

What happens as time increases? Clearly, the piston's energy will increase while its slope will decrease and the gas' energy will decrease while its slope will increase. This process will continue until the system reaches equilibrium as shown in Figure 28.4(b). At equilibrium, the slope for the piston becomes equal to the slope for the gas and the total entropy is maximized!

From this description, it sounds like the slope (i.e., dS/dU) is intimately connected to what we've been calling the temperature. Since heat flows from the system with the lower slope to the system with the higher slope, we can define **temperature** by identifying the inverse of the temperature $(1/T)$ as the derivative of the entropy with respect to the internal energy, when the volume is held constant.

$$\frac{1}{T} \equiv \frac{dS}{dU} \qquad \text{(at constant volume)}$$

With this definition, heat then flows from the hot system to the cold system until their temperatures are the same! In the next section, we will make the connection between this absolute temperature and the temperature scales you are more familiar with.

28.5   Temperature Scales

We have just defined the absolute temperature in terms of the change in entropy per unit energy of a system. With this definition we saw that heat flows from regions of higher temperature to regions of lower temperature until equilibrium is reached when all regions of the system are at the same temperature. For all systems we will discuss in this class, entropy increases with energy, so that the temperature we have defined is always positive. This temperature is an absolute temperature whose units are Kelvin (K).

While this definition of temperature is grounded in the fundamental concepts of entropy and energy, you may be wondering how we could use it to actually measure the temperature of a real system! We will answer this question in the next unit where we will use this definition of the temperature to derive the familiar **ideal gas law** in which this temperature is proportional to the product of the pressure and volume of the gas and inversely proportional to the number of atoms in the gas.

$$PV = NkT$$

Therefore, we could determine this temperature, for example, by measuring the pressure of a known quantity of the gas in a container of known volume.

To make the connection between this absolute Kelvin scale and the more familiar Celsius scale, we can compare them at the triple point of water, which corresponds to the temperature and pressure at which the three phases of water (solid, liquid, gas) can coexist in equilibrium. The temperature of the triple point of water is defined to be 273.16 K or 0.01° C. The magnitude of 1° C is defined to be equal to 1 K. Consequently, absolute zero (0 K) corresponds to −273.15° C. The Fahrenheit scale is defined at two points: the freezing point of water at 32° F and the boiling point of water at 212° F. Absolute zero (0 K) corresponds to −459.67° F.

Consequently, we can write down the conversions from one scale to another as follows:

Celsius to Kelvin: $T_K = T_C + 273.15$

Fahrenheit to Kelvin: $T_K = \dfrac{5}{9}\left(T_F + 459.67\right)$

Fahrenheit to Celsius: $T_C = \dfrac{5}{9}\left(T_F - 32°\right)$

28.6   Heat Capacity

We have just defined the temperature in terms of the slope of the entropy vs. energy plot as shown in Figure 28.5(a). Consequently, by taking the slope of this curve at every energy, we can generate the temperature *vs.* energy plot shown in Figure 28.5(b).

Note that the temperature of the gas increases with increasing energy, as you might guess. The rate of this increase (the slope of this curve) is used to define the heat capacity of the gas. Namely, we define the **heat capacity** of any object, C, as the amount of heat required to raise its temperature by 1 K.

$$C \equiv \frac{dQ}{dT}$$

For our stationary piston-gas system, the only energy change with temperature is the heat that has flowed from the gas to the piston. Consequently, the heat capacity at constant volume for the gas, C_V, is given directly from the slope of the curve in Figure 28.5(b).

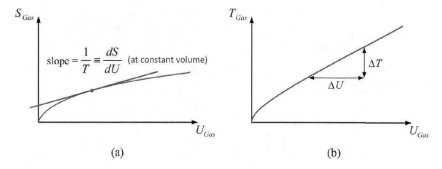

FIGURE 28.5 (a) Plot of entropy vs. energy for the stationary piston-gas system, identifying the slope of the curve as the inverse temperature at that point. (b) Plot of temperature vs. energy for the stationary piston-gas system, derived from the plot in part (a) of this figure.

$$C_V = \frac{dU}{dT} \to \frac{\Delta U}{\Delta T} \qquad \text{(constant volume)}$$

The heat capacity of an object is proportional to its mass; consequently, we can define the **specific heat capacity** for a given substance, c, as the heat capacity per unit mass.

$$c_V \equiv \frac{C_V}{m} = \frac{1}{m}\frac{dU}{dT} \to \frac{1}{m}\frac{\Delta U}{\Delta T} \qquad \text{(constant volume)}$$

The measured specific heat capacity at constant volume of helium gas over a wide range of temperatures is about 3.12 kJ/kg-K. Suppose the volume of the gas being heated is 1 m³, which corresponds to about 0.17 kg. If we were to add 1 kJ to this system, its temperature would rise by about 2 K.

$$\Delta T = \frac{1}{m}\frac{\Delta U}{c_V} = \frac{1}{(0.17\,\text{kg})}\frac{(1.0\,\text{kJ})}{(3.12\frac{\text{kJ}}{\text{kg·K}})} = 1.9\,\text{K}$$

Now, helium (He) is one of five stable monatomic noble gases that occur naturally, the others being neon (Ne), argon (Ar), krypton (Kr), and xenon (Xe). The measured specific heat capacity at constant volume for Ne, for example, is 0.618 kJ/kg-K, almost exactly $1/5$ that of He. Now, the atomic weight of Ne is almost exactly five times that of He. Similar relations hold for the rest of the stable monatomic noble gasses. If we now define the **molar heat capacity**, c_{mol}, as the heat capacity per mole of the substance,

$$c_{V\,mol} \equiv \frac{C_V}{n} = \frac{m}{n}c_V \qquad \text{(constant volume)}$$

we see from Table 28.1 that the molar heat capacities at constant volume of all the monatomic noble gases are equal to about 12.5 J/mol-K!

TABLE 28.1 Specific heats at constant volume of the stable monatomic noble gases.

	Atomic Weight	Specific Heat c_V (kJ / kg-K)	Molar Heat Capacity $c_{V\,mol}$ (J / mol-K)
Helium (He)	4.003	3.12	12.5
Neon (Ne)	20.1797	0.618	12.5
Argon (Ar)	39.948	0.312	12.5
Krypton (Kr)	83.80	0.15	12.5
Xenon (Xe)	131.29	0.095	12.5

Our calculation of specific heats does not carry over for gases that are not monatomic! For example, the molar heat capacities at constant volume of hydrogen (H_2) gas and nitrogen (N_2) gas are around 20 J/mol-K near room temperature. Molar heat capacities for solids such as copper (Cu) are typically around 25 J/mol-K. Since heat transfer is clearly connected to the internal structure of a substance, we will need to develop some microscopic models to be able to understand the heat capacities of materials that are not monatomic gases. Indeed, the study of such models will be the focus of the next unit.

28.7   Phases of Matter and Latent Heats

In the last section we discussed the heat capacity of a substance, the quantity that determines how much heat is required to raise its temperature. We know, however, that this concept does not capture the whole story. Sometimes, we can add heat and have the temperature not rise at all! Of course the situation described here is when we have a phase change, for example when a liquid becomes a solid or a gas, or when a solid becomes a liquid.

A simple example will illustrate this issue and will also show us the way forward. Suppose we put a pot containing 1 kg of water on the stove and turn on the burner. Heat flows from the flame into the pot holding the water, and from the pot into the water itself. If we look up the specific heat capacity of water ($c_{V\,water}$ = 4.2 kJ/kg-K), we can calculate that 4.2 kJ of heat will be required to raise the temperature by one Kelvin. If we know the initial temperature of the water, say for example its lowest temperature of 273.15 K, we can calculate that 420 kJ of heat is needed to raise the water's temperature all the way to the boiling point of 373.15 K.

As the heat continues to be added to the water, however, its temperature does not increase any further - instead the water starts to boil. To get some understanding of what is happening, we need to consider the molecular structure of water. In its liquid form, the water molecules are weakly bonded together. Therefore, to take a water molecule from its liquid state to its gaseous state, energy needs to be provided to break these bonds. While the water boils, all of the heat that flows into the water is used for breaking these bonds and turning the liquid into a gas, hence the temperature of the liquid water no longer changes. The energy required to turn 1 kg of a liquid into a gas is called the **latent heat of vaporization**.

$$L_v = \frac{\Delta Q_{liquid \to gas}}{m_{liquid}}$$

For water, this latent heat of vaporization is 2,257 kJ/kg. That is, 2,257 kJ of heat energy must be added to 1 kg of water at its boiling point to change the phase of this water from liquid to gas. Until this energy is added to the liquid, its temperature will not increase. A similar thing happens when a solid is heated to its melting point, and is referred to as the **latent heat of fusion**.

$$L_f = \frac{\Delta Q_{solid \to liquid}}{m_{solid}}$$

As heat flows into a melting block of ice its temperature is always 273.15 K, since the energy is used to convert the stronger bonds of the solid into the weaker bonds of the liquid. The latent heat of fusion for water is 334 kJ/kg.

Table 28.2 shows the latent heats of fusion and vaporization for various familiar materials. It is important to note that the same phenomenon happens in reverse when heat is removed from a system. In other words, heat must be removed in order for a gas to condense into a liquid, or for a liquid to freeze into a solid.

TABLE 28.2 Latent heats of familiar materials.

	$T_{M.P.}$ (K)	L_f (kJ / kg)	$T_{B.P.}$ (K)	L_v (kJ / kg)
Copper	1356	205	2839	4726
Gold	1336	62.8	3081	1701
Helium	-	-	4.2	21
Lead	600	24.7	2023	858
Mercury	234	11.3	630	296
Nitrogen	63	25.7	77.35	199
Oxygen	54.4	13.8	90.2	213
Water	273.15	333.5	373.15	2257

28.8 An Example

We will finish with an example that will let us practice using many of the ideas introduced in this unit. Suppose we have an insulated cup that contains 250 g of water having a temperature of 80° C. Fifty grams of ice whose temperature is 0° C is now dropped into the cup. After a long time, what will you find in the cup?

We will assume that the cup is sufficiently insulated such that no heat can leave or enter the system composed of the water and the ice. Therefore, any heat that flows into the ice must also flow out of the water. The heat required to turn ice at 0° C into water at 0° C is just equal to the product of the mass of the ice and the latent heat of fusion of water.

$$Q_{ice \to water} = m_{ice} L_{f,water} = (0.05\,\text{g})(334\,\text{kJ/kg}) = 16.7\,\text{kJ}$$

This heat must come from the surrounding water, so we can use the specific heat capacity of water to calculate the decrease in the temperature of the surrounding water as this heat is removed from it.

$$m_{water} c_{V,water} \Delta T = -16.7\,\text{kJ} \qquad \Rightarrow \qquad \Delta T = -15.9°$$

Therefore, the final temperature of the water is 64.1° C.

We now have 50 g of water at 0° C and 250 g of water at 64.1° C. Heat will now flow from the warmer water into the colder water until their temperatures are the same. What will be the final temperature of the water? To make this calculation, we simply set the amount of heat that flows out of the warmer water equal to the amount of heat that flows into the colder water.

$$-m_{water} c_{V,water} (T_f - 64.1°\,\text{C}) = m_{ice} c_{V,water} (T_f - 0°\,\text{C})$$

$$\Rightarrow \qquad T_f = (64.1°\,\text{C}) \left(\frac{m_{water}}{m_{water} + m_{ice}} \right)$$

Thus, we see that the final equilibrium condition is that the cup has 300 g of water in it at a temperature of 53.4° C.

MAIN POINTS

Work Done in Piston-Gas System

Applying the first law of thermodynamics ($\Delta U = W_{Net} + Q$) to the piston-gas system shown, we determined that the work done by the piston on the gas is given by:

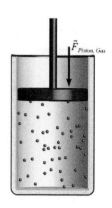

$$W_{Net} = -\int_{V_i}^{V_f} P\, dV$$

Temperature

We used the second law of thermodynamics that states that nature always flows to maximize the total entropy to determine the equilibrium equation for two materials in thermal contact and ultimately to define an absolute temperature scale.

$$\Delta S \geq 0$$

At equilibrium, the total entropy of the two systems will be a maximum and the derivative of the entropy with respect to the internal energy of each system will become the same for both systems.

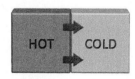

We defined the absolute temperature of a system as the inverse of the derivative of its entropy with respect to its internal energy.

$$\frac{1}{T} \equiv \frac{dS}{dU} \quad \text{(at constant volume)}$$

Heat Capacities and Latent Heats

The molar heat capacity of a substance was defined to be the amount of heat needed to raise the temperature of one mole of that substance by 1 K.

$$c_{V\,mol} = \frac{1}{n}\frac{\Delta Q}{\Delta T}$$

The latent heat of vaporization (fusion) was defined to be the energy needed to turn 1 kg of a liquid (solid) into a gas (liquid).

Latent Heat of Vaporization

$$L_v = \frac{\Delta Q_{liquid \rightarrow gas}}{m_{liquid}}$$

Latent Heat of Fusion

$$L_f = \frac{\Delta Q_{solid \rightarrow liquid}}{m_{solid}}$$

UNIT

29

IDEAL GAS

29.1 Overview

In this unit we will develop a model for monatomic gases that will allow us to calculate their molar heat capacities.

We start by considering a system composed of an insulated cylinder containing an ideal gas and a piston that is free to move as shown in Figure 29.1. We will find that by applying the equilibrium condition, namely that entropy is maximized, we will actually be able to derive the familiar ideal gas law: that the product of the pressure and volume of the gas is proportional to the product of the number of molecules of the gas and the absolute temperature.

$$PV = NkT$$

We will then calculate the pressure of an ideal gas which, when combined with the ideal gas law itself, will yield the important result that the average translational kinetic energy of an ideal gas molecule is proportional to the temperature of the gas!

This result will lead us to the more general equipartition relation, that the thermal energy in each degree of freedom of a molecule is equal to $kT/2$. We will then apply equipartition to our model for an ideal gas in order to calculate its molar heat capacity.

29.2 Entropy of an Ideal Gas

In our first unit on thermodynamics we introduced the concept of entropy, and the idea nature always moves to the configuration in which it is a maximum. We used this idea to show that two samples of a gas that are free to exchange volumes will come to equilibrium when their particle densities are the same. That is, when the pressures of the samples become equal.

In our second unit on thermodynamics we used the concept of entropy to understand why heat flows from warmer regions to colder ones, and even to define what we mean by temperature! We've come a long way basing everything on our initial concept of entropy. We aren't done yet, though; we will now use these same ideas to derive the familiar ideal gas law.

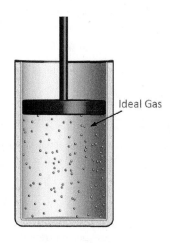

The insulated cylinder in Figure 29.1 is filled with a monatomic ideal gas. By "ideal", we mean that the atoms are so far apart that each one can freely move throughout the entire volume and that the potential energy of interactions between the atoms is negligible. Consequently, we can consider the internal energy of the gas to be simply given by the sum of the kinetic energies of the individual atoms.

FIGURE 29.1 An insulated cylinder containing an ideal gas and a piston that is free to move.

The system comes to equilibrium when the N molecules of the gas occupy a volume V. What is the entropy of this system? In the first unit of this thermodynamic section, we learned that at a fixed energy, the entropy of the gas is proportional to the logarithm of its volume since the number of states available to each particle is just proportional to the volume in which it can move.

Fixed Energy: $\Omega \propto V^N$ $\qquad \Rightarrow \qquad$ $S = k \ln(V^N) = kN \ln(V)$

How does this expression for the entropy change when the energy of the gas can change as well? It can be shown, using the properties of an ideal gas, that the entropy simply changes by the addition of a term that depends on energy, but not volume.

$$S = kN \ln(V) + f(U, N)$$

In the next section, we will use this expression for the entropy of the gas in order to derive the ideal gas law!

29.3 The Ideal Gas Law

In the last section, we introduced he expression for the entropy of an ideal gas whose volume and energy can both change.

$$S = kN \ln(V) + f(U, N)$$

We now need to justify this expression before we use it to derive the ideal gas law. For an ideal gas, the interactions between the particles have almost no effect on the average energy, so that the typical number of states of motion available to each particle per unit volume depends on the energy, but not on the density of the particles. Therefore, the number of states available to a particle is just the product of the volume and a term that just depends on energy. Remembering that the log of a product is the sum of the logs, we can see that the volume and energy dependence of the energy is captured by the simple expression shown above.

Now, at equilibrium, the entropy of the gas is maximized, so that the derivative of the entropy with respect to volume must be zero.

$$\frac{dS}{dV} = 0 \qquad \text{(equilibrium)}$$

Consequently, for small volume changes the combined change of the two terms in the entropy expression must be zero.

$$\frac{d}{dV}\left[kN \ln(V) + f(U, N)\right] = 0$$

We will now demonstrate that by just evaluating this expression we can obtain the ideal gas law!

If the piston were to move, the volume of the cylinder would change, clearly changing the first term. However, this change in volume would also result in work being done on the gas which would change its internal energy thereby also changing the second term! Consequently, we need to be careful that we evaluate dS/dV properly by taking partial derivatives with respect to both volume and internal energy.

$$\frac{dS}{dV} = \left(\frac{\partial S}{\partial V}\right)_U + \left(\frac{\partial S}{\partial U}\right)_V \frac{dU}{dV} = 0$$

The first term is simple to evaluate.

$$\left(\frac{\partial S}{\partial V}\right)_U = \frac{d}{dV}(kN \ln V) = \frac{kN}{V}$$

Looking at the second term, we see that one factor $(\partial S/\partial U)_V$ is just the inverse of the temperature that we defined in the last unit!

$$\left(\frac{\partial S}{\partial U}\right)_V = \frac{1}{T}$$

The remaining derivative in this term (dU/dV) can be determined since the change in internal energy of the gas is just equal to the work done on the gas by the piston!

$$dU = dW_{pisont,gas} = -PdV \qquad \Rightarrow \qquad \frac{dU}{dV} = -P$$

We can now just put all these pieces together to arrive at the familiar **ideal gas law**.

$$PV = NkT$$

Amazing! Starting with our general expression for the entropy of an ideal gas whose volume and energy can both change, we were able to derive the ideal gas law by requiring that the entropy is maximized at equilibrium!

29.4   Pressure of an Ideal Gas

We will now use some elements of classical mechanics (namely, the momentum and kinetic energy of individual molecules) to determine the pressure of an ideal gas.

Figure 29.2 shows a single molecule as it bounces of a wall of a cubical container whose sides have length L. Only the x component of the molecule's velocity is changed by this collision, and, assuming an elastic collision, the magnitude of this change is easy to find.

$$\Delta v_x = 2v_x$$

As the molecule bounces back and forth between the wall at $x = 0$ and $x = L$, its speed in the x direction does not change, so that the time that elapses before it hits the same wall again is just the distance that it travels divided by this speed.

$$\Delta t = \frac{2L}{v_x}$$

Now the average force exerted on the wall by the molecule is just the change in momentum divided by the change in time.

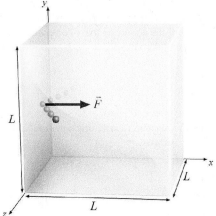

FIGURE 29.2 An ideal gas molecule collides with the wall of a container. The momentum of the molecule is changed by the force $\vec{F}$ the wall exerts on the molecule.

$$F_{avg} = \frac{\Delta p}{\Delta t}$$

The change in momentum is determined by the change in velocity.

$$\Delta p = m\Delta v_x = 2mv_x$$

Putting this altogether, we see that the average force exerted on the wall by the molecule is proportional to the square of the x component of its velocity.

$$F_{avg} = m\frac{v_x^2}{L}$$

We can convert this average force to a pressure by simply dividing by the area of the wall.

$$P = \frac{F_{avg}}{Area} = \frac{mv_x^2}{L^3} = \frac{mv_x^2}{V}$$

Since the single molecule we picked is only one of a very large number of molecules, the total pressure on the wall due to all of the atoms can be found by multiplying our expression by the number of molecules and then replacing the square of the x component of the velocity of the single molecule by the average value of all of the molecules.

$$P = N\frac{m\langle v_x^2 \rangle}{V}$$

Since there is no preferred direction in our box, the average value of the square of any component of the velocity must be the same. Consequently, we can replace the average value of the square of the x component of the velocity by $1/3$ of the average value of the square of the velocity.

$$P = \frac{1}{3}N\frac{m\langle v^2 \rangle}{V}$$

Finally, we can rewrite this equation in terms of the average translational kinetic energy of the molecules.

$$P = \frac{1}{3}N\frac{m\langle v^2 \rangle}{V} = \frac{2}{3}N\frac{\langle \frac{1}{2}mv^2 \rangle}{V} = \frac{2N}{3V}\langle K \rangle_{trans}$$

Of course in a real gas the molecules also collide with each other. That's how they end up with random directions and similar energies. However, the pressure is just the same as we calculated for molecules that don't collide, because the pressure comes completely from the molecules hitting the surface. How they happened to get there doesn't matter, just how often and how hard they hit.

In the next section we will combine this result for the pressure with the ideal gas law to obtain a connection between the internal energy and the temperature of an ideal gas.

29.5   Equipartition

In the third section of this unit, we derived the ideal gas law ($PV = NkT$) from the second law of thermodynamics, and in the last section we used our knowledge of mechanics to arrive at a different expression for the product of the pressure and volume of an ideal gas.

$$PV = \frac{2N}{3}\langle K \rangle_{trans}$$

We can combine these equations to obtain a relationship between the temperature of an ideal gas and its average translational kinetic energy.

$$\langle K \rangle_{trans} = \frac{3}{2}kT$$

We'll now consider the special case of a monatomic gas where the only kinetic energy is translational in order to make three important observations.

First, we see that the average kinetic energy of a gas atom is proportional to the temperature of the gas itself. This is not surprising since we tend to think of temperature and energy as being related, but it's reassuring to see that we can arrive at this result from first principles. The constant of proportionality, k, is called **Boltzmann's constant** and it has been well measured experimentally.

$$k = 1.381 \times 10^{-23} \text{ J / K}$$

Second, this remarkable result relates the dynamics of an *individual microscopic molecule* to a measureable property of a *large macroscopic system*. For example, consider a balloon filled with helium gas at 20° C. We can calculate the average kinetic energy of a helium atom in this balloon to be about 6×10^{-21} J. Knowing the mass of a helium atom ($M = (4 \text{ amu})(1.66 \times 10^{-27} \text{ kg / amu})$), we can then then determine the average speed of these molecules to be about 1,400 m/s.

Third, we've shown that the average kinetic energy per atom in these gases is independent of the mass of the atoms! This is the first piece of a much more general relation, called **equipartition**, showing that thermal energy goes equally into various types of modes. Remarkably, this relation holds for a variety of modes other than the simple translational motions, including rotational kinetic energy. When equipartition applies, we can determine the average thermal energy by simply counting the "degrees of freedom" (*d.o.f.*), the number of independent modes that can get thermal energy.

$$\langle U \rangle / d.o.f. = \frac{1}{2}kT$$

For the kinetic energies we looked at, the thermal energy was $(3/2)NkT$, corresponding to $3N$ degrees of freedom, since each of the N molecules could move in three independent directions. Consequently, each degree of freedom got $(1/2)NkT$, on average. We will discuss other degrees of freedom when we discuss diatomic molecules in the next unit.

29.6   Molar Heat Capacity at Constant Volume

We saw in the last unit that the heat capacity for one mole of a fixed volume of any monatomic ideal gas, such as helium, neon, argon, krypton and xenon, is always about 12.5 J/mol-K. We are now in a position to explain this remarkable fact.

Consider one mole of a monatomic ideal gas in a container of fixed volume. We will calculate the heat capacity of this gas sample by determining the amount of heat required to raise its temperature by 1 K.

$$C_V \equiv \left(\frac{dQ}{dT} \right)_V$$

Since the volume is being held constant, no work is being done by the gas so that the first law of thermodynamics tells that the change in internal energy is just equal to the heat that is added to the gas. Consequently, the heat capacity is just given by the change in internal energy with temperature.

$$C_V = \frac{dU}{dT}$$

But we just found an expression for the internal energy of a monatomic ideal gas $(U = (3/2)NkT)$, so we can now write a simple expression for the heat capacity of our gas.

$$C_V = \frac{3}{2} Nk$$

Since one mole contains **Avogadro's number** of molecules $(N_A = 6.022 \times 10^{23}$ molecules$)$, and since we know the value of Boltzmann's constant, we see that we predict the molar heat capacity at constant volume for an ideal monatomic gas to be 12.47 J/mol-K, which is exactly what the experiments measure.

$$c_{V,mol} \equiv \frac{C_V}{n} = \frac{3Nk}{2n} = \frac{3(6.022 \times 10^{23})(1.381 \times 10^{-23} \text{ J/K})}{2(1 \text{ mol})} = 12.47 \frac{\text{J}}{\text{mol} \cdot \text{K}}$$

Wow! Using arguments based on entropy and classical mechanics we have in one fell swoop correctly calculated the molar heat capacity of all monatomic gases, and have thereby validated our expression for the average kinetic energy of the molecules of this gas. In the next unit, we will extend our analysis to diatomic molecules which have more than three degrees of freedom.

29.7 Molar Heat Capacity at Constant Pressure

As long as we are thinking about heat capacity, let's consider another variation on this problem that will be of interest to us in the next two units as we make our way to a study of heat engines. Instead of keeping the volume of the gas fixed as we add heat, which is what we discussed in the previous section, we will allow the volume to increase as we add the heat in such a way that the pressure stays constant.

We once again start with the first law of thermodynamics that relates heat to internal energy and work for the gas system.

$$dU = dQ - P\,dV$$

Since the heat capacity is defined as the rate of change of heat with temperature $(C \equiv dQ/dT)$, we can find the heat capacity at constant pressure by simply differentiating this equation.

$$C_P \equiv \left(\frac{dQ}{dT}\right)_P = \frac{dU}{dT} + P\frac{dV}{dT}$$

The first term in this expression is just the heat capacity at constant volume that we found in the previous section.

$$C_P = C_V + P\frac{dV}{dT}$$

The second term in this expression can be found by differentiating the ideal gas law (remember P is being held constant).

$$PV = NkT \qquad \Rightarrow \qquad P\frac{dV}{dT} = Nk$$

If we evaluate this expression for one mole of a gas, we predict that the molar heat capacity at constant pressure should be larger than the molar heat capacity at constant volume by an amount equal to the product of Avogadro's number and the Boltzmann constant, which is equal to 8.31 J/mol-K.

$$c_{P,mol} \equiv \frac{C_P}{n} \qquad \Rightarrow \qquad c_{P,mol} - c_{V,mol} = N_A k = 8.31\,\frac{J}{mol \cdot K}$$

Table 29.1 shows the measured values for the heat capacities at room temperature for several gases showing excellent agreement for this difference for the noble gases and very good agreement for the other gases that are not monatomic.

Note that the heat capacity at constant pressure is always bigger than the heat capacity at constant volume. That is to say, if the volume of the gas is allowed to increase as heat is added to the gas, the temperature does not increase as much as it does if the same amount heat is added to the gas when the volume is fixed. We can understand this result in terms of the first law of thermodynamics, which tells us that the internal energy of a gas increases when heat is added to it, but that it decreases when the gas expands and does work.

$$dU = dQ - P\,dV$$

Since the temperature of an ideal gas is proportional to its internal energy, we can see that the temperature of a gas will go down when it does work to expand unless heat is added to it. This is why the air you let out if a bicycle tire feels colder than the surrounding air – since the air does work pushing aside the atmosphere as it expands when it leaves the tire, its temperature must decrease.

TABLE 29.1 Measured molar heat capacities (in J / mol-K) at constant volume and constant pressure for some gases.

	$c_{P\,mol}$	$c_{V\,mol}$	$c_{V\,mol} - c_{P\,mol}$
He Helium	20.786	12.472	8.31
Ne Neon	20.786	12.472	8.31
Ar Argon	20.786	12.472	8.31
Kr Krypton	20.786	12.472	8.31
Xe Xenon	20.786	12.472	8.31
N_2 Nitrogen	29.12	20.8	8.32
O_2 Oxygen	29.38	21.0	8.38
CO_2 Carbon Dioxide	36.94	28.46	8.48

MAIN POINTS

Ideal Gas Law

The system, composed of an insulated cylinder containing an ideal gas and a piston that is free to move, achieves equilibrium at some volume V, where entropy is maximized. Setting $dS / dV = 0$, we obtain the ideal gas law.

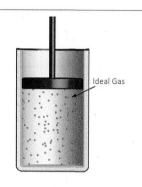

Ideal Gas

Ideal Gas Law

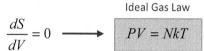

$$\frac{dS}{dV} = 0 \longrightarrow \boxed{PV = NkT}$$

Equipartition

Applying some elements of mechanics, we obtained a relation between the pressure, volume, and average translational kinetic energy of a monatomic ideal gas. From the result that the average kinetic energy per atom is independent of the mass of the atom, we generalized our monatomic ideal gas result to obtain the equipartition relation: that the thermal energy in each degree of freedom (*d.o.f.*) of a molecule is equal to $(1/2)kT$.

Average Translational Kinetic Energy per Atom for an Ideal Gas

$$\left\langle K \right\rangle_{trans} = \frac{3}{2} kT$$

Equipartition

$$\left\langle U \right\rangle / d.o.f. = \frac{1}{2} kT$$

Molar Heat Capacities

Using the general first law of thermodynamics and the specific relation between the energy of an ideal monatomic gas and its temperature, we obtained a relation for the molar heat cpacity at constant volume for an ideal monatomic gas.

$$c_{V\,mol} = \frac{3}{2} N_A k = 12.47 \frac{J}{mol \cdot K}$$

Using the first law of thermodynamics and the ideal gas law, we obtained a relation for the difference between the molar heat capacities at constant pressure and at constant volume.

$$c_{P\,mol} - c_{V\,mol} = N_A k = 8.31 \frac{J}{mol \cdot K}$$

30

EQUIPARTITION, HEAT CAPACITY AND CONDUCTION

30.1 Overview

In this unit, we will continue our development of models that we can use to calculate the molar heat capacities of diatomic ideal gases and of solids.

We will introduce first a model for diatomic ideal gases that includes additional rotational degrees of freedom that will lead to a molar heat capacity that is larger than that of monatomic ideal gases.

We will then introduce a model for solids that includes additional vibrational degrees of freedom, representing the bonds by connecting the atoms with little springs. Using our knowledge of simple harmonic motion from mechanics we will be able to determine the additional degrees of freedom in this model and therefore be able to calculate the molar heat capacity of solids.

We will close by simply introducing the parameterizations used to describe two processes of practical interest, namely those of thermal expansion and heat conduction.

30.2   Additional Degrees of Freedom: Diatomic Molecules

In the last unit, we found that the average kinetic energy per atom of a monatomic gas was equal to $(3/2)kT$, independent of the mass of the atom! This result was our first example of a more general relation, called equipartition, showing that thermal energy goes equally into various types of modes. For the monatomic gas, the only modes available are the translational kinetic energy modes. We said then that when equipartition applies, we can determine the average thermal energy by simply counting the "degrees of freedom", the number of independent modes that can get thermal energy.

Figure 30.1 shows the two extra degrees of freedom available to a diatomic model. Namely, the molecule can rotate around two distinct axes, each perpendicular to the axis of the molecule. These two extra degrees of freedom then can be added to the three translational degrees of freedom to make a total of five degrees of freedom for the diatomic molecule. Each of these degrees of freedom gets $(1/2)kT$ average energy under a broad range of conditions. For these molecules equipartition predicts that the total average kinetic energy due to both translations and rotations should be given by $(5/2)kT$. We will soon see that this is in fact the case. We won't prove it, but it turns out that equipartition holds for all sorts of modes of motion so long as the temperature is high enough to neglect quantum mechanical effects.

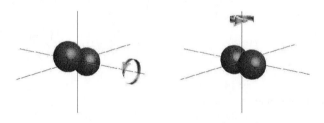

FIGURE 30.1 A diatomic molecule can rotate about two independent axes, each perpendicular to the axis of the molecule.

We just argued that diatomic molecules have two extra degrees of freedom, for a total of five, because they can rotate around two distinct axes perpendicular to the axis of the molecule. Therefore, the internal energy of a mole of diatomic gas at a given temperature will be bigger than that of a monatomic gas at the same temperature and consequently, we expect the molar heat capacity for an ideal diatomic gas to be a factor of $5/3$ bigger than the molar heat capacity of an ideal monatomic gas.

$$U_{diatomic} = \frac{5}{2}NkT \qquad \Rightarrow \qquad (c_{V\,mol})_{diatomic} = \frac{5}{2}N_A k$$

$$(c_{V\,mol})_{diatomic} = \frac{5}{3}(c_{V\,mol})_{monatomic}$$

Looking back at Table 29.1 that shows the molar heat capacities for several molecules we see that once again our prediction ($c_{v,\,diatomic} = (5/3)(12.47) = 20.8$ J/K) agrees well with the experimental results. Despite the simplicity of our model, equipartition does an excellent job in explaining the heat capacity of these gasses.

30.3   Polyatomic Molecules

Assuming equipartition, we have calculated the molar heat capacities at constant volume for monatomic and diatomic ideal gases and generally found good agreement with the measured values at room temperature as shown in Table 29.1.

We have attributed the differences in c_V between the monatomic and diatomic gases to their differences in internal energy, the number of degrees of freedom that are available. Note that the difference between c_P and c_V, though, is essentially the same for all gases shown. We can understand this result in that this difference is due solely to the work done on heating at constant pressure, which we were able to calculate by differentiating the ideal gas law.

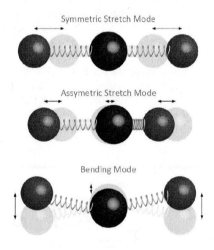

FIGURE 30.2 The three vibrational modes of the linear molecule, CO_2.

Figure 30.2 shows a representation of the CO_2 molecule, the one entry in the table that we have not yet discussed. We have represented the bonds as little springs. This representation allows for vibrational modes that we did not model in the diatomic gases. In fact there ·are there three   distinct   vibrational   modes. Consequently, we expect these additional modes to increase the molar heat capacity above that of the diatomic molecules, as shown. The exact amount of increase depends on the details of the molecular binding, on the energy required for each mode. We will not carry this analysis any further here due to these complications.

In the next section, however, we will consider a similar model for solids that we can use to calculate molar heat capacities.

30.4   A Model for Solids

So far, we have developed a model for ideal gases that allows us to make predictions about the internal energy of the gas under various conditions, which in turn allows us to calculate the heat capacity of the gas. We will now introduce a simple model that will allow us to do the same thing for certain solids.

In particular, we will adopt the model shown in Figure 30.3 in which the atoms of the solid are point masses that are bound together by little springs. This model is sometimes called the "bedspring" model for obvious reasons.

The internal energy of a monatomic ideal gas was simply due to the average kinetic energy of the gas atoms as they move throughout the volume. In the same way, the internal energy of the atoms in our model solid will in part be due to the kinetic energy of the atoms as they jiggle around their equilibrium positions. Just as was true for the ideal gas, equipartition says the average kinetic energy of each atom in the solid should be $(3/2)kT$ since it is free to vibrate in three dimensions.

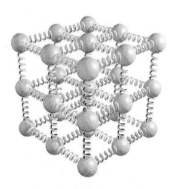

FIGURE 30.3 The bedspring model of a solid in which the bonds between atoms are represented as springs.

So how do the springs in this model contribute to the internal energy? When we studied the simple harmonic motion of a mass on a spring in mechanics, we found that as the mass oscillated back & forth the kinetic energy of the mass and the potential energy of the spring always added up to a constant total mechanical energy even though the magnitude of each one oscillated. Since both the kinetic and potential energy oscillate harmonically between zero and the total energy, the average of each one is just half of the total. In other words, *the average kinetic energy of the mass is equal to the average potential energy of the spring*, and each one on average accounts for half of the total mechanical energy of the system.

Applying this argument to our model of a solid as a system of masses and springs leads us to conclude that the total average internal energy for each atom in the solid is just twice its average kinetic energy, and that the total internal energy for a solid containing N atoms is just N times the average internal energy of a single atom.

TABLE 30.1 Molar heat capacities of some common solids.

	c_{mol} (J/mol-K)
Aluminum	24.2
Copper	24.5
Gold	25.4
Iron	25.1
Magnesium	24.1
Lead	26.4
Cadmium	26.0
Bismuth	25.7
Silver	24.9

$$\left\langle U_{per\ atom} \right\rangle = 3kT \qquad \Rightarrow \qquad U_{Total} = 3NkT$$

The volume change as we add heat to a solid is usually quite small, so that any work done is negligible. Consequently, we can approximate any heat transfer by the change of the internal energy of the solid. Therefore, we expect that the molar heat capacity of our solid should be twice that of an ideal monatomic gas.

$$c_{mol} \equiv \frac{1}{n}\frac{dQ}{dT} = \frac{1}{n}\frac{dU}{dT} = \frac{N}{n}3k = 3N_A k = 24.9\ N_A k = 8.31\ \frac{J}{mol \cdot K}$$

Table 30.1 shows measured values of the molar heat capacity near room temperature for some common solids, and we can see that our model actually does very well.

30.5 Quantum Effects and Equipartition

We have just seen that using models based on equipartition, we have been able to predict the heat capacities of many gasses and solids near room temperature. Equipartition is, at root, a classical explanation. The underlying theory of our world is quantum mechanical; the energies of the microstates we have been talking about are not continuous, but rather quantized. Indeed, as temperature is decreased, the available energy for the microparticles is reduced and eventually will not be sufficient to support all modes required by equipartition. Consequently, we expect the heat capacity to eventually decrease with decreasing temperature.

We can actually see this quantum effect at room temperature for a few materials. For example, the molar heat capacities of beryllium and diamond are 16.4 J/K and 6.1 J/K at room temperature, rather than the familiar 25 J/K. Figure 30.4 shows the molar heat capacities of diamond, beryllium and silver as a function of temperature.

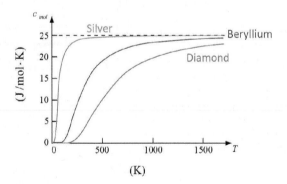

FIGURE 30.4 The molar heat capacity of silver, beryllium, and diamond as a function of temperature.

At high temperatures, the molar heat capacities of all solids approach the classical prediction of 25 J/K, but the approach to this limit depends on the detailed quantum structure of the solid.

30.6   Thermal Expansion

We had to distinguish between C_V and C_P for gases, because at fixed pressure these gases expanded on heating enough to do quite a bit of work, $P\Delta V$ to be precise. Our solids expand or contract very little, so we don't need to consider that volume-change term except in precise heat capacity measurements. However, that small thermal expansion/contraction of solids is very important in many engineering problems. For example, if thermal expansion is not considered in constructing large systems (think railroad tracks, bridges) there could be disastrous consequences!

In general, materials are found to expand when their temperature increases. Linear expansion of a material is characterized by its **coefficient of linear expansion**, α.

$$\alpha \equiv \frac{1}{L}\frac{dL}{dT}$$

The units for α are K^{-1}, and its value depends on the internal structure of the material and varies quite a lot as seen in Table 30.2.

Of course, all dimensions expand equally, so we can also define a **coefficient of volume expansion**, β.

$$\beta \equiv \frac{1}{V}\frac{dV}{dT}$$

TABLE 30.2 Coefficients of expansion for some common materials.

Solid	$\alpha\,(K^{-1})$
Diamond	1.2×10^{-6}
Graphite	7.9×10^{-6}
Glass	9×10^{-6}
Steel	11×10^{-6}
Copper	17×10^{-6}
Aluminum	24×10^{-6}
Ice	51×10^{-6}

Writing the volume as a product of the dimensions of a box, we can see that $\beta = 3\alpha$.

$$V = L_1 L_2 L_3 \qquad \Rightarrow \qquad \beta = \frac{1}{V}\frac{dV}{dT} = \sum_{i=1}^{i=3}\frac{1}{L_i}\frac{\partial L_i}{\partial T} = 3\alpha$$

Before we leave this topic, it is important to note that thermal expansion is *not* universal; a number of materials actually contract upon heating, the most notable example being water between 0°C and 4°C. The maximum density for water occurs at 4°C. Consequently, cooling water below 4°C will cause the water to expand! This expansion is important. As water cools in a lake below 4°C, it becomes less dense and rises to the top. This expansion then explains why lakes freeze first at the surface.

30.7   Heat Conduction

We've seen that thermal energy flows to maximize the entropy S, going from regions of high-temperature to regions of low temperature. We haven't said anything yet about *how* it flows. In solids, those thermal vibrations consist of tiny sound waves, traveling around bouncing off each other and off irregularities in the material. Looking at any region, more of these waves will be coming in from the high-temperature side than from the low-temperature side. The net flow of energy will be proportional to the difference in the temperatures, at least for small differences.

We characterize this heat flow in terms of κ, the **thermal conductivity** of the material. Namely, κ is equal to minus the heat flow per unit area divided by the temperature gradient.

$$\frac{dQ/dt}{A} = -\kappa \frac{dT}{dx}$$

The minus sign is needed because the heat flows from high temperatures to low temperatures, opposite to the gradient in temperature. For example, if the heat gradient is positive (the temperature is increasing with increasing x), the heat flow will be negative (heat will flow from positive x to negative x).

The units for κ are W/m-K and the values vary considerably as shown in Table 30.3. We can see that electrical conductors have much higher conductivities than electrical insulators. For example, a one meter length of #6 cooper wire (area $= 13\,\text{mm}^2$) connecting an ice bath to a steam bath will have a heat flow of about one-half watt!

In gases the mechanism for thermal conductivity (energy transfers from collisions) is different from that in solids, but the heat flow can still be characterized by this quantity, κ.

TABLE 30.3 Thermal conductivities for some common materials.

	κ(W/m-K)
Air at 27°C	0.026
Oak	0.15
Ice	0.592
Water at 27°C	0.609
Glass	0.7–0.9
Steel	46
Iron	80.4
Aluminum	237
Gold	318
Copper	401

MAIN POINTS

Diatomic Molecules

We introduced a model for diatomic gases in which the bond between the atoms is modeled as a rigid rod. The molecule then has two additional degrees of freedom corresponding to rotations about two axes that are perpendicular to the rod. Using equipartition, we obtained a prediction for the molar heat capacity at constant volume of diatomic gases that agrees well with the measurements for many diatomic gases at room temperature.

$$U_{diatomic} = \frac{5}{2} NkT$$

$$(c_{V\,mol})_{diatomic} = \frac{5}{2} N_A k = 20.8 \frac{J}{mol \cdot K}$$

Ideal Solids

We introduced the bedspring model for solids in which the bonds between atoms were modeled as springs. Using properties of the simple harmonic oscillator and equipartition, we obtained the prediction that the molar heat capacity of solids at room temperature should be twice the molar heat capacity at constant volume of of an ideal monatomic gas.

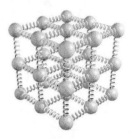

$$U = 3NkT$$

$$c_{mol} = 3N_A k = 24.9 \frac{J}{mol \cdot K}$$

Limitations to Our Calculations

Our calculations of molar heat capacities all were derived from expressions for the internal energy of molecules which were obtained by simply using equipartition, that thermal energy goes equally into all available modes. This assumption works well at high temperatures for a wide variety of substances, but begins to fail as the temperature is lowered, due to the detailed quantum structure of each substance.

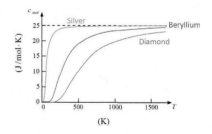

31

HEAT ENGINES

31.1 Overview

In this unit, we will introduce the concept of the heat engine. That is, we will look at systems in which the working substance of an engine, usually a gas, extracts heat from a hot reservoir, does some useful work and then exhausts any remaining heat to a cold reservoir.

We will begin by defining the efficiency of the engine as the ratio of the work done by the engine to the heat extracted from the hot reservoir. We will find that the second law of thermodynamics places an upper limit on this efficiency that is determined by the temperature difference of the reservoirs.

We will then introduce the Stirling cycle as our standard example of a heat engine. This cycle consists of two isothermic transitions (i.e., at constant temperature) and two isochoric transitions (i.e., at constant volume). We will calculate the efficiency of this engine when the working substance is an ideal gas and will compare it to maximum efficiency allowed by the second law of thermodynamics.

31.2   Heat Engines

You are probably already familiar with the concept of an engine: it consumes energy and in return makes something useful happen. An internal combustion engine uses gasoline or diesel fuel to turn a drive shaft to make cars or trucks move; the steam engine of an old locomotive uses coal as a fuel to turn the locomotive's wheels; and a nuclear power plant uses uranium as fuel to turn generators that make electricity.

These engines are clearly quite different, yet they do share some common features. In this course we will focus on heat engines, namely those that convert heat into mechanical work. All heat engines are driven by a temperature difference. The idea is to extract some work as heat flows from a hot region to a cold region. In our car engine, gasoline is mixed with air and burned, causing the gas on one side of a piston to be hotter than the gas of the other side. This results in a pressure difference between the hot and cold side, causing the piston to move and the engine to do useful work.

Another defining feature of a heat engine is that its operation must be *cyclical*. In other words, it keeps doing the same thing over and over again, each time with the same result. In the car engine, as the piston moves back to its original position, the hot gas that just caused the piston to move must be replaced by a new batch of colder air mixed with gasoline. In this way, the whole process can repeat again.

31.3   Turning Heat into Work

Since a difference in temperature is key to the operation of all heat engines, it is common to represent the operation of an engine through one of its cycles using the diagram shown in Figure 31.1 in which heat flows from a hot reservoir through the working substance of the engine, usually a gas, to a cold reservoir. The engine converts some of this heat to mechanical work, usually by the expansion and compression of the gas during a complete cycle.

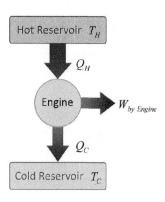

We can apply the first law of thermodynamics to the engine to determine that the heat that flows from the hot reservoir into the engine must be equal to the sum of the work done by the engine and the heat that flows out of the engine into the cold reservoir.

FIGURE 31.1 Diagram representing a heat engine. Heat is extracted from a hot reservoir, some of this energy is converted to work, and the rest is exhausted to the cold reservoir.

$$Q_H = W + Q_C$$

In order to keep the temperature of the hot reservoir constant, energy must be continually added to it. For example, if a certain amount of energy is extracted each cycle, then an equivalent amount of energy must also be supplied to the reservoir to maintain its temperature. This is the energy cost to drive the engine. The engine then does work. Consequently, it is natural to define the **efficiency** of the engine as the ratio of the work done by the engine to the heat that has to be extracted from the hot reservoir.

$$\varepsilon \equiv \frac{W}{Q_H}$$

In the next section, we will show that the second law of thermodynamics places a limit on how large this efficiency can be. We will find that this limit is determined by the temperature difference of the reservoirs.

31.4   Efficiency of a Heat Engine

We have just defined the efficiency of a heat engine to be the ratio of the work done by the engine to the heat that was extracted from the hot reservoir. We can rewrite this expression in terms of the heat transfers with the two reservoirs by applying the first law of thermodynamics to the engine.

$$Q_H = W + Q_C \qquad \Rightarrow \qquad \varepsilon = 1 - \frac{Q_C}{Q_H}$$

Consequently, we see that a 100% efficient heat engine would convert all of the heat extracted from the hot reservoir into work. That is, no heat would flow into the cold reservoir. We shall soon see that this process cannot happen since it would violate the second law of thermodynamics.

To see how this works, let's look at the entropy change of the engine plus the reservoirs during one cycle. Since the working substance of the engine is returned to its initial state at the completion of one cycle, its entropy does not change!

$$\Delta S_{engine} = 0$$

Now, the entropies of each reservoir, however, do change. We can determine these entropy changes from our definition of temperature.

$$\frac{1}{T} \equiv \frac{\Delta S}{\Delta U} \qquad \Rightarrow \qquad \Delta S = \frac{\Delta U}{T}$$

Since the reservoirs do no work, the first law of thermodynamics tells us that the change in the internal energy of each reservoir is just equal to the heat that flows into the reservoir.

$$\Delta U_H = -Q_H \qquad \text{and} \qquad \Delta U_C = Q_C$$

Consequently, the change in the entropy of each reservoir is equal to the ratio of the heat flowing into the reservoir to the temperature of the reservoir.

$$\Delta S_H = -\frac{Q_H}{T_H} \qquad\qquad \text{and} \qquad\qquad \Delta S_C = \frac{Q_C}{T_C}$$

The total entropy change of the engine plus reservoirs is just equal to the sum of the entropy changes of the reservoirs.

$$\Delta S_{Total} = \Delta S_{engine} + \Delta S_H + \Delta S_C = 0 - \frac{Q_H}{T_H} + \frac{Q_C}{T_C}$$

Now, the second law of thermodynamics says this total entropy change must always be greater than or equal to zero! Consequently, we see that the ratio of the heat transfers from the reservoirs must always be greater than or equal to the ratio of the temperatures of the reservoirs.

$$\frac{Q_C}{T_C} \geq \frac{Q_H}{T_H} \qquad\qquad \Rightarrow \qquad\qquad \frac{Q_C}{Q_H} \geq \frac{T_C}{T_H}$$

Going back to our expression for the efficiency of a heat engine, we see that this efficiency can never be greater than one minus the ratio of the temperatures of the reservoirs.

$$\varepsilon = 1 - \frac{Q_C}{Q_H} \leq 1 - \frac{T_C}{T_H}$$

This maximum efficiency is usually referred to as the **Carnot efficiency**, after Sadi Carnot, a French physicist and engineer of the 19th century.

$$\varepsilon_{Carnot} = 1 - \frac{T_C}{T_H}$$

31.5 Quasi-static Processes and PV Diagrams

We have just seen that there is a limit for the efficiency of a heat engine that is proportional to the temperature difference of the reservoirs.

$$\varepsilon \leq 1 - \frac{T_C}{T_H}$$

For example, a heat engine operating between reservoirs of boiling water (100°C = 373 K) and freezing water (0°C = 273 K) has a maximum efficiency of 27%. We would now like to look at more realistic engines and calculate their efficiencies to see how they compare to this Carnot limit.

In order to make this calculation, we will use the **quasi-static approximation**, that all changes are done slowly enough so that the gas is essentially at equilibrium at all times, meaning that it has a well-defined pressure, temperature, volume, and entropy at all times.

Making this quasi-static approximation allows us to represent the cycle on a *PV* diagram. That is, at any time during the cycle, the gas will have a well-defined pressure and volume so that the state of the gas at that time can be represented as a unique point in the pressure-volume plane. Figure 31.2 shows the *PV* diagram for a particular engine, the Stirling cycle, that we will study in more detail in the next two sections.

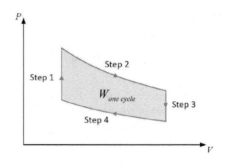

For now, we will simply note that the work done by the engine during one cycle is captured in this diagram as the area enclosed by the curve. That is, the work done by the gas is always equal to

$$W_{by\ gas} = \int_{cycle} P\ dV$$

FIGURE 31.2 *PV* diagram for one cycle of a Stirling engine. Steps 2 and 4 are isothermal transitions.

so that we can simply add up the work done by the gas during each step. No work is done in step 1 or step 3 since the volume does not change. The engine does positive work in step 2 equal to the area under the curve as the gas expands. The engine does negative work in step 4 equal to the area under the curve as it is compressed. We will now look at this cycle in more detail.

31.6   The Stirling Cycle

In the *PV* diagram for the Stirling engine shown in Figure 31.2, we see that steps 2 and 4 are **isothermal transitions** (i.e., constant temperature) made at the temperatures of the hot and cold reservoirs and that steps 1 and 3 are **isochoric transitions** (i.e., constant volume).

We will now describe each step in the cycle in more detail. Each step is illustrated in Figure 31.3. Let's look at each step in the cycle in turn.

In step 1 the volume of the cylinder is held fixed. The gas, which starts at the temperature of the cold reservoir, is put in contact with the hot reservoir. Heat flows from the reservoir into the gas, increasing its pressure, until the temperature of the gas is the same as the reservoir.

In step 2 the gas is still in contact with the hot reservoir so that its temperature is fixed. The volume of the cylinder is increased as the piston rises which means the pressure drops.

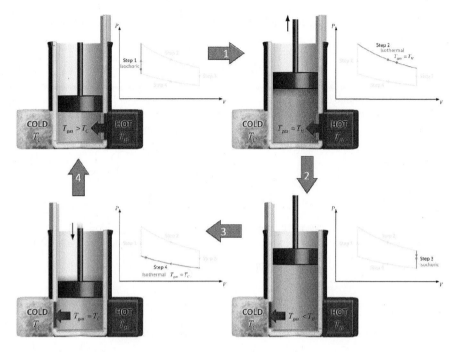

FIGURE 31.3 The four transitions that comprise the Stirling cycle. Steps 1 and 3 are isochoric (constant volume), and steps 2 and 4 are isothermal (constant temperature).

In step 3 the volume of the cylinder is held fixed. The gas, which starts at temperature of the hot reservoir, is put in contact with the cold reservoir. Heat flows from the gas into the reservoir, decreasing the pressure, until the temperature of the gas is the same as the reservoir.

In step 4 the gas is still in contact with the cold thermal reservoir so that its temperature is fixed. The volume of the cylinder is decreased as the piston falls which means the pressure increases. In the next section, we will calculate the efficiency of this engine when the working substance is an ideal gas.

31.7   Efficiency of the Stirling Engine with an Ideal Gas

In order to determine the efficiency of the Stirling engine, we need to calculate the work done by the gas for a known heat input from the hot reservoir. To make this calculation, we will assume the working substance is an ideal gas, so that we can use the ideal gas law to relate the pressure, volume and temperature of the gas at any point during the cycle.

$$PV = NkT$$

In step 1, heat flows from the hot reservoir into the gas. This heat is just equal to the product of the heat capacity at constant volume and the temperature difference of the reservoirs.

$$Q_1 = C_V \left(T_H - T_C \right)$$

In step 2, the volume expands and the gas does work. We can calculate this work since we know how the pressure changes with volume during this transition since we have assumed it is an ideal gas. Integrating over the change in volume gives us a result that is proportional to the product of the temperature and the log of the ratio of the volumes.

$$W_2 = \int_{step\ 2} P\, dV = \int_{step\ 2} \frac{NkT_H}{V} dV = NkT_H \ln \left(\frac{V_{big}}{V_{small}} \right)$$

Since the temperature does not change, the internal energy of the gas does not change, and the first law tells us that the heat that flows into the gas is the same as the work done by the gas.

$$Q_2 = W_2$$

In step 3, heat flows from the gas into the cold reservoir. Since the temperature change is just the opposite of step 1, the heat flow is also just the opposite.

$$Q_3 = -Q_1$$

In step 4, the volume is compressed and work is done on the gas. We can calculate this work since we know how the pressure changes with volume during this transition, since we have assumed the working substance of the engine is an ideal gas. Once again, integrating over the volume gives us the result that the work done is proportional to the product of the temperature and the log of the ratio of the volumes.

$$W_4 = \int_{step\ 4} P\, dV = \int_{step\ 4} \frac{NkT_C}{V} dV = -NkT_C \ln \left(\frac{V_{big}}{V_{small}} \right)$$

As in step 2, the heat flow has the same magnitude as the work.

$$Q_4 = W_4$$

Putting this all together, the total work done by the gas is proportional to the product of the temperature difference and the logarithm of the ratio of the volumes.

$$W = W_2 + W_4 = Nk \left(T_H - T_C \right) \ln \left(\frac{V_{big}}{V_{small}} \right)$$

The total heat extracted from the hot reservoir is equal to the sum of the heat from steps 1 and 2.

$$Q_{in} = Q_1 + Q_2 = C_V(T_H - T_C) + NkT_H \ln\left(\frac{V_{big}}{V_{small}}\right)$$

We now substitute our expressions for the work done by the gas and the heat added to the gas into our efficiency relation.

$$\varepsilon \equiv \frac{W}{Q_{in}} = \frac{Nk(T_H - T_C)\ln\left(\frac{V_{big}}{V_{small}}\right)}{C_V(T_H - T_C) + NkT_H \ln\left(\frac{V_{big}}{V_{small}}\right)}$$

We can rewrite this expression by factoring out the Carnot efficiency explicitly so that we can see how this Stirling efficiency compares to the maximum Carnot efficiency.

$$\varepsilon = \frac{T_H - T_C}{T_H}\left[\frac{Nk \ln\left(\frac{V_{big}}{V_{small}}\right)}{C_V\left(\frac{T_H - T_C}{T_H}\right) + Nk \ln\left(\frac{V_{big}}{V_{small}}\right)}\right]$$

The fraction that multiplies the Carnot efficiency ($(T_H - T_C)/T_H$), even though it looks complicated, must be less than one since the denominator has to be bigger than the numerator. Therefore, the efficiency of the Stirling engine is indeed less than the Carnot efficiency.

For example, if we assume a doubling of volume for a monatomic ideal gas between reservoirs at the temperatures of boiling and freezing water, we obtain an efficiency for the Stirling engine of 17% while the Carnot efficiency for these temperatures is 27%.

31.8 A Stirling Engine

We have just calculated the efficiency of a Stirling engine and found it to be less than the Carnot efficiency. In the next unit, we will develop a general expression for the efficiency of a heat engine in terms of the increase in the entropy of the engine and its environment that will explain exactly why the Stirling efficiency is less than the Carnot efficiency.

We close with a final note of interest to those of you who may be wondering how to actually build a Stirling engine. Since it is not practical to repeatedly heat and then cool the same walls of a cylinder, the usual design uses some method to move the gas back and forth between the hot part and the cold part of the cylinder so that the temperatures of these parts never have to change. In the design shown in Figure 31.5, the gas is moved between the hot bottom and cold top of the cylinder by a displacer. Notice that the displacer is smaller than the cylinder so it does not compress the gas, it just moves the gas around. We will discuss real-world designs more in the next unit.

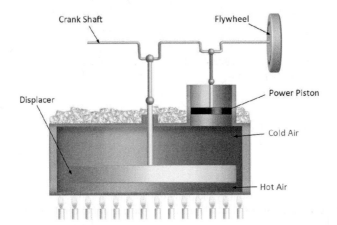

FIGURE 31.4 A Stirling engine: the gas is moved between the hot and cold reservoirs by a displacer.

MAIN POINTS

Heat Engines

We introduced the concept of a heat engine in which heat is extracted from a hot reservoir, does work, and then exhausts heat to a cold reservoir. We defined the efficiency of the engine to be the ratio of the work done by the engine to the heat that was extracted from the hot reservoir. We applied the second law of thermodynamics to the engine plus reservoirs to determine that there is a maximum efficiency, called the Carnot efficiency, which is equal to the ratio of the temperature difference between the reservoirs to the temperature of the hot reservoir.

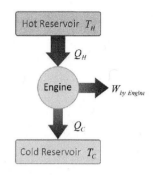

Engine Efficiency

$$\varepsilon \equiv \frac{W_{by\ Engine}}{Q_H} \leq \frac{T_H - T_C}{T_H}$$

Carnot Efficiency

$$\varepsilon_{Carnot} = \frac{T_H - T_C}{T_H}$$

Quasi-static Approximation

In order to calculate the efficiencies of heat engines, we restricted the possible processes in the engine cycle to those that are quasi-static, meaning that all changes are done slowly enough so that the working substance of the engine is essentially at equilibrium at all times. Consequently, we could represent any engine cycle on a PV diagram in which the work done by the engine is represented as the area enclosed by the transitions that make up the cycle.

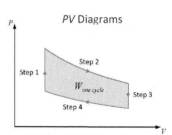

Example: Stirling Cycle

We introduced the Stirling cycle as our standard example of a heat engine. This cycle consists of two isothermic transitions (constant temperature) and two isochoric transitions (constant volume). We calculated the efficiency of this engine, using an ideal gas as the working substance and found that the efficiency was indeed less than the Carnot efficiency.

$$\varepsilon_{Stirling} = \frac{T_H - T_C}{T_H} \left[\frac{Nk \ln\left(\dfrac{V_{big}}{V_{small}} \right)}{C_V \left(\dfrac{T_H - T_C}{T_H} \right) + Nk \ln\left(\dfrac{V_{big}}{V_{small}} \right)} \right]$$

UNIT

32

REVERSIBLE PROCESSES

32.1 Overview

In this unit, we will conclude our study of heat engines. Our main focus will be on the connection between the efficiency of a heat engine and the nature of the processes that are used in the cycle.

We will begin by determining that there are only two kinds of processes that are reversible, those being quasi-static adiabatic and isothermal processes. All other processes are irreversible; that is, the total entropy of the system and its environment must increase in these processes. Indeed, we will discover that the efficiency of a heat engine is reduced from its maximum Carnot efficiency by an amount that is proportional to the increase in the total entropy of the system and its environment during a cycle.

Finally, we will look at two important devices, refrigerators and heat pumps, which operate by running a heat engine in reverse. Namely, in these devices, heat is caused to flow from the cold reservoir to the hot reservoir by doing work on the engine.

32.2 Heat Engine Efficiencies

In the last unit, we introduced the idea of a heat engine as represented in Figure 31.1 and defined its efficiency as the ratio of the work done during a cycle to the heat extracted from the hot reservoir.

$$\varepsilon \equiv \frac{W}{Q_H}$$

We used the second law of thermodynamics to determine the maximum possible efficiency of any heat engine in terms of the temperatures of the hot and cold reservoirs.

$$\varepsilon_{Carnot} \equiv \varepsilon_{max} = 1 - \frac{T_C}{T_H} = \frac{T_H - T_C}{T_H}$$

This maximum efficiency, the Carnot efficiency, is realized when the change in the total entropy of the engine and its environment is equal to zero as the engine goes through one complete cycle.

We introduced the Stirling engine whose cycle is represented in Figure 31.2 and calculated its efficiency when the working substance was an ideal gas. We found that its efficiency was indeed less than the Carnot efficiency, indicating that the change in the total entropy of this engine and its environment must be greater than zero. That is, irreversible processes must exist somewhere in the Stirling cycle.

In order to understand this loss in efficiency, we will need to look at reversible processes in more detail, which we will now do.

32.3 Reversible Processes I

Our immediate task is to identify all of the quasi-static reversible processes. We start with something familiar. Figure 32.1 shows a well-insulated cylinder with a piston that is free to move. We start by assuming that the gas in the cylinder is in equilibrium. Now, just as we did when we derived the ideal gas law three units ago, we make the claim that a small quasi-static change in the volume of the gas produces no change in the total entropy of the gas and its environment. Certainly there is no change in the entropy of the environment since it is completely insulated from the gas. However, there is also no change in the entropy of the gas, since it remains in equilibrium during the quasi-static volume change; that is, entropy is maximum at equilibrium, meaning that the derivative of the entropy of the gas with respect to volume is zero.

$$\frac{dS_{gas}}{dV} = 0$$

Consequently, a small change in the volume produces no change in the entropy. This process, in which there is no heat flow between the engine and its environment, is called an **adiabatic process**.

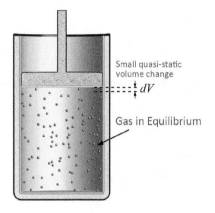

FIGURE 32.1 An insulated cylinder containing a gas in equilibrium and a piston that is free to move.

Note that the argument we just made only works because the environment is insulated from the gas. If this were not the case, heat could flow between the gas and the environment, and we have no guarantee that the entropy of the environment would not change.

We have just discovered that *quasi-static adiabatic processes are reversible*. On the next slide we will investigate non-adiabatic quasi-static transitions to determine if there are any conditions under which these can be reversible.

32.4   Reversible Processes II

Figure 32.2 shows a *PV* diagram for a quasi-static transition in which both the volume and the pressure of the gas change by small amounts *dV* and *dP*. Since the total entropy change between the initial and final states does not depend on the path between those states, we will calculate this change using the alternate path also shown. Namely, the volume of the gas is first increased adiabatically to its final value and then the pressure is increased to its final value, keeping the volume constant.

As we just saw, there is no change in the total entropy during the adiabatic step. To find the entropy change of the gas during the second step we will need to find the change in its internal energy.

$$dS_{gas,isochoric} = \frac{dU_{gas,isochoric}}{T_{gas}} = 0$$

Now, since there is no change in the volume during this step, no work is done by the gas, so

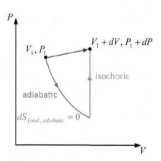

FIGURE 32.2 *PV* diagram illustrating two paths connecting the initial state (V_1, P_1) to a final state in which the volume and pressure are each changed by a small amount (*dV* and *dP*).

that the change in its energy is just equal to the heat that flows into it from the environment.

$$dU_{gas,isochoric} = dW_{gas,isochoric} + dQ_{gas} = 0 + dQ_{gas}$$

Consequently, the entropy of the gas must increase during this step by an amount equal to the heat that flows into it divided its temperature.

$$dS_{gas,isochoric} = \frac{dQ_{gas}}{T_{gas}} = 0$$

Now the heat that flows into the gas during this step is equal to the heat that left the environment since no work was done.

$$dQ_{environment} = -dQ_{gas}$$

Therefore, the entropy of the environment must decrease during this step by an amount equal to the heat that flows into the gas divided by the temperature of the environment.

$$dS_{environment} = -\frac{dQ_{gas}}{T_{environment}}$$

Therefore, the change in the total entropy of the gas plus its environment is equal to the product of the heat that flows from the environment to the gas and the difference between the inverse of the temperatures of the gas and its environment.

$$dS_{Total} = dS_{gas,isochoric} + dS_{environment,isochoric} = \frac{dQ_{gas}}{T_{gas}} - \frac{dQ_{gas}}{T_{environment}}$$

$$dS_{Total} = dQ_{gas}\left(\frac{1}{T_{gas}} - \frac{1}{T_{environment}}\right)$$

There are only *two ways* to make this expression equal to zero, which means that there are only *two processes* that will leave the total entropy unchanged. The first is when the heat flow between the gas and the environment is zero (i.e., a quasi-static adiabatic process).

$$dQ_{gas} = 0 \qquad \text{(Adiabatic Process)}$$

The second is when the temperature of the gas and the environment is the same (i.e., a quasi-static isothermal process).

$$T_{gas} = T_{environment} \qquad \text{(Isothermal Process)}$$

Note that we have called the system here a gas, but have in no way used any property of a gas to obtain this important result. This result holds for all systems! Armed with this knowledge of reversible processes, we can now address the efficiency of the Stirling engine, which we will do next.

32.5   Entropy Increase of a Stirling Engine

Figure 32.3 shows the *PV* diagram for the Stirling cycle. steps 2 and 4 are quasi-static *isothermal* transitions which we now know are *reversible* processes. Steps 1 and 3 however are quasi-static *constant volume* transitions which we now know must be *irreversible* since they are neither adiabatic nor isothermal. Consequently, the total entropy of the engine and its environment *must increase* during these two steps.

We will now calculate the increase in total entropy for the Stirling cycle. We don't have to worry about the change in entropy of the gas itself because we know it comes back to its starting state after each cycle. Therefore, we just need to calculate the changes of the reservoirs.

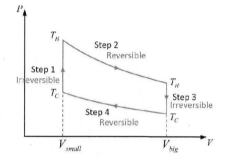

FIGURE 32.3 Steps 1 and 3 of the Stirling cycle are irreversible transitions while steps 2 and 4 are reversible processes.

During step 1, the change in the entropy of the reservoir is equal to minus the ratio of the heat that flows into the gas to the temperature of the reservoir.

$$\Delta S_{1,\,Hot\,Res} = -\frac{Q_1}{T_H}$$

Likewise in step 3, where heat flows into the *cold* reservoir.

$$\Delta S_{3,\,Cold\,Res} = -\frac{Q_3}{T_C}$$

If we can assume the heat capacity at constant volume, C_V, is constant between T_C and T_H, our calculation simplifies. Namely, in this case, the heat flow is just equal to the product of C_V and the temperature difference between the reservoirs.

$$Q_1 = -Q_3 = C_V\left(T_H - T_C\right)$$

We can now calculate the total entropy change by summing the entropy changes of the reservoirs during steps 1 and 3.

$$\Delta S_{Total} = \Delta S_{1,\,Hot\,Res} + \Delta S_{3,\,Cold\,Res}$$

$$\Delta S_{Total} = C_V \left(T_H - T_C \right) \left(-\frac{1}{T_H} + \frac{1}{T_C} \right)$$

It is this increase in entropy that is responsible for the efficiency of the Stirling engine being less than the Carnot efficiency. We will now use this entropy increase to calculate the efficiency of the Stirling engine.

32.6   Efficiency from Entropy Increase

We have just determined the increase in entropy during one cycle of a Stirling engine. We can now use this result to determine the efficiency of the engine. Namely, the efficiency of any engine is just equal to one minus the ratio of the heat flow to the cold reservoir to that from the hot reservoir.

$$\varepsilon \equiv \frac{W_{by\ engine}}{Q_H} = 1 - \frac{Q_C}{Q_H}$$

Now, in general, since the change in entropy of the gas is zero over one cycle, the change in the total entropy is just equal to the change in the entropy of the reservoirs.

$$\Delta S_{Total} = \Delta S_{1,\ Hot\ Res} + \Delta S_{3,\ Cold\ Res} = \frac{Q_C}{T_C} + \frac{(-Q_H)}{T_H}$$

If we now multiply this expression by the ratio of the temperature of the cold reservoir to the heat that flows out of the hot reservoir, we can identify a term which is equal to the efficiency.

$$\frac{T_C}{Q_H} \Delta S_{Total} = \frac{Q_C}{Q_H} \frac{T_C}{T_H} = \left(1 - \varepsilon \right) - \frac{T_C}{T_H}$$

We can rewrite this expression to obtain a totally general expression for the efficiency which illustrates that the reduction from the Carnot efficiency for any engine is proportional to the change in the total entropy.

$$\varepsilon = 1 - \frac{T_C}{T_H} - \Delta S_{Total} \frac{T_C}{Q_H} = \varepsilon_{Carnot} - \Delta S_{Total} \frac{T_C}{Q_H}$$

We can recover our expression for the Stirling efficiency for an ideal gas from this general result if we just substitute in the expressions for ΔS_{Total} above and for Q_H from the last unit when we assumed the working substance in the Stirling engine was an ideal gas.

$$\mathcal{E}_{Stirling} = \mathcal{E}_{Carnot} \left[\frac{Nk \ln\left(\dfrac{V_{big}}{V_{small}}\right)}{C_V \left(\dfrac{T_H - T_C}{T_H}\right) + Nk \ln\left(\dfrac{V_{big}}{V_{small}}\right)} \right]$$

32.7 The Carnot Cycle

We have just seen that the entropy increases during steps 1 and 3 of the Stirling cycle are responsible for its efficiency being less than that of the Carnot cycle. This entropy analysis, though, tells us exactly how to create a Carnot engine! We simply need to replace steps 1 and 3 of the Stirling cycle by adiabatic processes, since these processes are reversible and do not increase the total entropy.

We will now describe each step in the cycle in more detail. Each step is illustrated in Figure 32.4. Let's look at each step in the cycle in turn. In step 1 the entire cylinder is insulated so that no heat can flow between the gas and its environment, and the piston is moved down so that the volume of the gas decreases. Since work is being done on the gas by the piston, the internal energy of the gas, and therefore its temperature, increases.

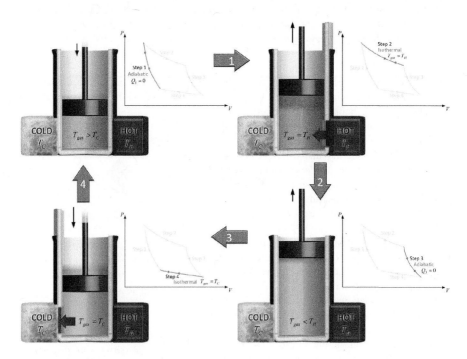

FIGURE 32.4 The four transitions that comprise the Carnot cycle. Steps 1 and 3 are adiabatic, and steps 2 and 4 are isothermal.

Once the temperature of the gas is the same as that of the hot reservoir we start step 2, in which the insulation is removed and the cylinder is placed in contact with the hot reservoir and the gas is allowed to expand at constant temperature.

In step 3 the cylinder is insulated again, and as the volume expands further, both the pressure and the temperature decrease.

Once the temperature of the gas in the same as that of the cold reservoir we start step 4, in which the insulation is removed and the cylinder is placed in contact with the cold reservoir and the gas is allowed to contract at constant temperature. At this point the process can repeat.

The Carnot engine is the most efficient engine between two temperatures, but it is not a practical engine, partly due to the fact that those adiabatic processes require big volume changes when the ratio of the temperatures is large enough to obtain a high efficiency. The more compact Stirling engines are used in many applications.

32.8   Running a Heat Engine in Reverse

To this point, we have focused on heat engines in which the natural heat flow from hot to cold can produce work. If, on the other hand, we do work on the system, we can cause heat to flow from cold to hot. Indeed, two very important devices, refrigerators and heat pumps, operate on this very principle. The schematic that illustrates this idea is shown in Figure 32.5.

Let's start with a refrigerator. Here, the cold reservoir is the inside where we keep the milk. The hot reservoir is the air surrounding the refrigerator, and the work is supplied by an electric motor hooked up to the refrigerator's heat engine, which is often called a compressor. When the refrigerator is running, the heat engine removes heat from the inside of the fridge to keep the temperature inside constant and cold, and moves this heat to the surrounding air. This is why the air behind a fridge is usually warmer than the rest of the air in the room.

A heat pump operates on the same principle as a refrigerator except the focus is on the heat flowing into the hot side of the engine rather than the heat flowing out of the cold side. For example, we may want to use the heat being transferred from the cold reservoir into the hot reservoir to heat a house. In this case the hot reservoir is the air in the house, and the cold reservoir is the air outside the house, or perhaps the water in a nearby lake.

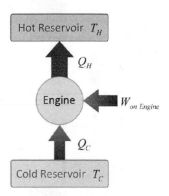

FIGURE 32.5 Work is done on the engine in order to extract heat from the cold reservoir to transfer some heat to the hot reservoir. Running the engine in reverse is used in refrigerators and heat pumps.

Figure 32.6(a) shows the Carnot cycle for a heat engine. What would the Carnot cycle look like for a refrigerator operating between the same temperatures? Well, to make this diagram, all we have to do is to reverse the arrows on each process, as shown in Figure 32.6(b)! We will now discuss the figures of merit for these devices.

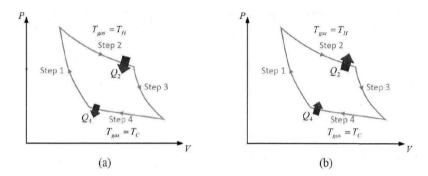

FIGURE 32.6 Carnot cycles for a heat engine (a) and a refrigerator (b).

32.9 Efficiency and Coefficient of Performance

We have defined the efficiency of a heat engine as the ratio of the work done to the heat extracted from the hot reservoir. The corresponding quantity for refrigerators and heat pumps is called the coefficient of performance. Since refrigerators and heat pumps have different objectives, though, we must define this coefficient differently for each device.

For a refrigerator, the coefficient of performance is defined to be the ratio of the heat it removes from the cold reservoir to the work done on the engine to make this happen.

$$COP_{refrigerator} = \frac{Q_C}{W_{on\ Engine}}$$

For a heat pump, the coefficient of performance is defined to be the ratio of the heat it adds to the hot reservoir to the work done on the engine to make this happen.

$$COP_{heat\ pump} = \frac{Q_H}{W_{on\ Engine}}$$

We can use conservation of energy $(Q_C + W_{on\ Engine} = Q_H)$ to express these coefficients of performance in terms of the ratio of the heat flows to the reservoirs.

$$COP_{refrigerator} = \frac{1}{\dfrac{Q_H}{Q_C} - 1}$$

$$COP_{heat\ pump} = \frac{1}{1 - \dfrac{Q_C}{Q_H}}$$

As with heat engines, the best performance is obtained when the total entropy change is zero. Therefore, the maximum coefficient of performance is obtained with a Carnot cycle. We can use our previous result for the Carnot cycle to obtain expressions for the maximum coefficient of performance in terms of the ratios of the temperatures of the reservoirs.

$$COP_{refrigerator,\ \text{max}} = \frac{1}{\dfrac{T_H}{T_C} - 1}$$

$$COP_{heat\ pump,\ \text{max}} = \frac{1}{1 - \dfrac{T_C}{T_H}}$$

Let's try a few numbers. For a refrigerator whose inside temperature is 5°C $(T_C = 278 \text{ K})$ in a kitchen whose temperature is 25°C $(T_C = 298 \text{ K})$, we find that the COP is nearly 14. This means that an efficient refrigerator can move much more energy from the cold side to the hot side than it takes to run it.

For a heat pump that keeps a house at 20°C $(T_C = 293 \text{ K})$ by transferring heat from a nearby lake whose temperature is 5°C $(T_C = 278 \text{ K})$, we find that the COP is close to 20. In other words, you would get almost twenty times more heat into the house using this heat pump, as you would by running the same amount of energy through an electric heater. In practice, real heat pumps cannot achieve this theoretical maximum COP, but do have still impressive COPs of around 5.

32.10   Adiabatic Processes for an Ideal Gas

The PV diagram for the Carnot cycle is shown in Figure 32.6. We have been able to derive most of our important results for heat engines without needing to know the exact functional relationship between P and V in the individual processes. However, if we want to calculate the work done during a cycle, for example, we definitely need to know P as a function of V. This functional relationship depends on the working substance of the engine. We will close this unit with a derivation of this relationship for adiabatic processes of an ideal gas.

We already know this functional relationship for isothermal processes. Three units ago, we derived the ideal gas law, that the product of P and V is proportional to the temperature.

$$PV = NkT$$

Therefore, for isothermal processes, the product of P and V is a constant.

To determine the relationship between P and V for adiabatic processes, we return to the example we used to derive the ideal gas law. We start from our expression for the entropy of an ideal gas when both the volume and the energy can change.

$$S = kN \ln(V) + f(U, N)$$

We know that during an adiabatic process the change in entropy of the gas is zero since the adiabatic process is reversible and the gas is insulated from its environment. However, during an adiabatic process, since both the volume and the energy of the gas change, we need to express dS in terms of partial derivatives of S with respect to volume and energy.

$$dS = \frac{\partial S}{\partial V} dV + \frac{\partial S}{\partial U} dU$$

We can evaluate the partial derivative of entropy with respect to the volume directly from our expression for the entropy.

$$\frac{\partial S}{\partial V} = k \frac{N}{V}$$

We can evaluate the partial derivative of the entropy with respect to the energy directly from our definition of temperature!

$$\frac{\partial S}{\partial U} = \frac{1}{T}$$

Therefore,

$$dS = kN \frac{dV}{V} + \frac{dU}{T}$$

Now, for an ideal gas, we can express dU in terms of dT and α defined to be one-half the number of degrees of freedom.

$$dU = C_V \, dT = \alpha kN \, dT$$

Therefore,

$$dS = kN \frac{dV}{V} + \alpha kN \frac{dT}{T}$$

Integrating this expression and setting the result equal to zero since adiabatic processes are reversible, we obtain the important result that for an adiabatic process for an ideal gas, VT^{α} is a constant!

$$\int \frac{dV}{V} + \alpha \int \frac{dT}{T} = 0$$

$$\alpha \ln\left(\frac{T_2}{T_1}\right) + \ln\left(\frac{V_2}{V_1}\right) = 0$$

Therefore,

$$\left(\frac{T_2}{T_1}\right)^{\alpha} = \frac{V_1}{V_2} \qquad \Rightarrow \qquad V_1 T_1^{\alpha} = V_2 T_2^{\alpha}$$

We can now use the ideal gas law to write the temperature in terms of the pressure and volume to obtain the equivalent form that PV^{γ} is a constant where γ is equal to C_P / C_V, which is also equal to $(\alpha + 1)/\alpha$.

$$V_1^{1/\alpha} T_1 = V_2^{1/\alpha} T_2 \qquad \Rightarrow \qquad V_1^{\gamma-1} T_1 = V_2^{\gamma-1} T_2$$

$$T \propto PV \qquad \Rightarrow \qquad P_1 V_1^{\gamma} = P_2 V_2^{\gamma}$$

There you have it! For an adiabatic transition of an ideal gas PV^{γ} is a constant. This result explains why the adiabatic transitions are steeper than the isothermal transitions in the PV diagram.

MAIN POINTS

Reversible Processes

We derived a general expression for the change in the total entropy of a system and its environment during any quasi-static process. We found that this change in this entropy was equal to the product of the heat that flows from the environment to the system and the difference of the inverse of the temperatures of the system and the environment. Therefore, there are only two possible reversible quasi-static processes. These are adiabatic processes and isothermal processes.

Quasi-Static Transition

$$dS_{Total} = dQ_{gas}\left(\frac{1}{T_{gas}} - \frac{1}{T_{environment}}\right)$$

Heat Engine Efficiencies and Entorpy

We determined that the efficiency of any heat engine is always less than the Carnot efficiency by an amount that is proportional to the increase in entropy of the system plus its environment. Thus, reductions from the maximum possible efficiency of a heat engine are due to the presence of irreversible processes in the cycle.

$$\varepsilon = \varepsilon_{Carnot} - \Delta S_{Total}\frac{T_C}{Q_H}$$

Refrigerators and Heat Pumps

We investigated two practical uses for running heat engines in reverse (i.e., by doing work on the engine we can make heat flow from cold to hot). In a refrigerator, work is supplied by an electric motor to remove heat from the cold reservoir (the inside of the refrigerator) to the hot reservoir (the surrounding air in the room). In a heat pump, the same principle is used, but the focus is on the heat flowing into the hot side of the engine rather than the heat flowing out of the cold side. The figures of merit (the coefficients of performance, COPs) for these devices are different because they have different purposes.

$$COP_{refrigerator} \equiv \frac{Q_C}{W_{on\ Engine}}$$

$$COP_{heat\ pump} \equiv \frac{Q_H}{W_{on\ Engine}}$$

APPENDIX A: NUMERICAL DATA

Some Fundamental Physical Constants*

Avogadro's number	N_A	$6.0221415(10) \times 10^{23}$ particles/mol
Coulomb constant	k	$8.987551788... \times 10^9$ N·m^2/C^2
Electron rest mass	m_e	$9.10938215(45) \times 10^{-31}$ kg
Elementary charge	e	$1.602176487(40) \times 10^{-19}$ C
Gravitational constant	G	$6.67428(67) \times 10^{-11}$ N·m^2/kg^2
Neutron rest mass	m_n	$1.674927211(84) \times 10^{-27}$ kg
Permeability constant	μ_o	$4\pi \times 10^{-7}$ N/A^2
Permittivity constant	ε_o	$8.85418781... \times 10^{-12}$ C^2/(N·m^2)
Proton rest mass	m_p	$1.672621637(83) \times 10^{-27}$ kg
Speed of light in a vacuum	c	299,792,458 m/s

*The values for these constants may be found on the Internet at http://physics.nist.gov/cuu/Constants/index.html. The numbers in the parenthesis represent the uncertainties in the last two digits. For example, the number $6.67428(67)$ equals 6.67428 ± 0.00067. Values without parenthesis represent exact values without uncertainties.

Astronomical Data*

Earth	
Mass	5.97×10^{24} kg
Radius	6.37×10^6 m
Distance from Sun†	1.496×10^{11} m
Moon	
Mass	7.35×10^{22} kg
Radius	1.737×10^6 m
Period	27.32 days
Distance from Earth†	3.844×10^8 m
Acceleration of gravity at surface	1.62 m/s^2
Sun	
Mass	1.99×10^{30} kg
Radius	6.96×10^8 m

*Data for our solar-system can be found on the Internet at http://nssdc.gsfc.nasa.gov/planetary/planetfact.html.
†Center to center.

APPENDIX B: SI UNITS

Base Units*

Meter (m)	The *meter* is the length of the path traveled by light in vacuum during a time interval of $1/299,792,458$ of a second.
Kilogram (kg)	The *kilogram* is the unit of mass; it is equal to the mass of the international prototype of the kilogram.
Second (s)	The *second* is the duration of 9,192,631,770 periods of the radiation corresponding to the transition between the two hyperfine levels of the ground state of the cesium 133 atom.
Ampere (A)	The *ampere* is that constant current that, if maintained in two straight parallel conductors of infinite length, of negligible circular cross-section, and placed 1 meter apart in vacuum, would produce between these conductors a force equal to 2×10^{-7} newton per meter of length.
Kelvin (K)	The *kelvin* is the fraction $1/273.16$ of the thermodynamic temperature of the triple point of water.
Candela (cd)	The *candela* is the luminous intensity, in a given direction, of a source that emits monochromatic radiation of frequency 540×10^{12} hertz and that has a radiant intensity in that direction of $1/683$ watt per steradian.
Mole (mol)	The *mole* is the amount of substance of a system that contains as many elementary entities as there are atoms in 0.012 kilogram of carbon 12.

*These definitions are found on the Internet at http://physics.nist.gov/cuu/Units/current.html

Derived Units

Force	newton (N)	$1\,\text{N} = 1\,\text{kg}\cdot\text{m/s}^2$
Work (Energy)	joule (J)	$1\,\text{J} = 1\,\text{N}\cdot\text{m}$
Power	watt (W)	$1\,\text{W} = 1\,\text{J/s}$
Frequency	hertz (Hz)	$1\,\text{Hz} = 1\,\text{cycle/s}$
Pressure	pascal (Pa)	$1\,\text{Pa} = 1\,\text{N/m}^2$
Charge	coulomb (C)	$1\,\text{C} = 1\,\text{A}\cdot\text{s}$
Potential	volt (V)	$1\,\text{V} = 1\,\text{J/C}$
Capacitance	farad (F)	$1\,\text{F} = 1\,\text{C/V}$
Current	ampere (A)	$1\,\text{A} = 1\,\text{C/s}$
Resistance	ohm (Ω)	$1\,\Omega = 1\,\text{V/A}$
Magnetic field	tesla (T)	$1\,\text{T} = 1\,\text{N/(A}\cdot\text{m)}$
Magnetic flux	weber (Wb)	$1\,\text{Wb} = 1\,\text{T}\cdot\text{m}^2$
Inductance	henry (H)	$1\,\text{H} = 1\,\text{J/A}^2$

APPENDIX C: CONVERSION FACTORS

Length
  1 m = 3.281 ft = 39.37 in
* 1 in = 2.54 cm
* 1 ft = 12 in = 30.48 cm
  1 mi = 5280 ft = 1.609 km
  1 km = 0.6214 mi

Time
* 1 h = 60 min = 3600 s
* 1 day = 24 h = 1440 min = 86,400 s
  1 yr = 365.25 days = 3.156×10^7 s

Angular Measurement
* π rad = 180° = ½ rev
  1 rad = 57.30° = 0.1592 rev
  1° = 0.01745 rad

Speed
  1 km/h = 0.2778 m/s = 0.6215 mi/h
  1 mi/h = 0.4470 m/s = 1.609 km/h

Acceleration
  9.8 m/s^2 = 32.2 ft/s^2

Force
  1 N = 0.2248 lb
* 1 lb = 4.448222 N
  A 1-kg mass weighs 2.203 lb,
    where g = 9.80 m/s^s

Pressure
* 1 Pa = 1 N/m^2
* 1 atm = 1.01325×10^5 Pa
  1 atm = 760 mmHg = 29.9 inHg
  1 bar = 10^5 Pa
  1 lb/in^2 = 6.895×10^3 Pa

Energy
  1 J = 0.7376 ft · lb
* 1 cal = 4.184 J
  1 kW · h = 3.6×10^6 J
  1 eV = 1.602×10^{-19} J

Power
* 1 W = 1 J/s
  1 W = 0.7376 ft · lb/s
  1 W = 1.341×10^{-3} hp
  1 hp = 550 ft · lb/s = 745.7 W

*Exact values.

INDEX